Computational Methods in Structural Engineering

Computational Methods in Structural Engineering

Guest Editors

Manolis Georgioudakis
Vagelis Plevris
Mahdi Kioumarsi

MDPI

Basel • Beijing • Wuhan • Barcelona • Belgrade • Novi Sad • Cluj • Manchester

Guest Editors

Manolis Georgioudakis
School of Civil Engineering
National Technical University of Athens
Athens
Greece

Vagelis Plevris
Department of Civil and Environmental Engineering
Qatar University
Doha
Qatar

Mahdi Kioumarsi
Department of Built Environment
Oslo Metropolitan University
Oslo
Norway

Editorial Office
MDPI AG
Grosspeteranlage 5
4052 Basel, Switzerland

This is a reprint of the Special Issue, published open access by the journal *Computation* (ISSN 2079-3197), freely accessible at: https://www.mdpi.com/journal/computation/special_issues/1T85OR9SSP.

For citation purposes, cite each article independently as indicated on the article page online and as indicated below:

Lastname, A.A.; Lastname, B.B. Article Title. *Journal Name* **Year**, *Volume Number*, Page Range.

ISBN 978-3-7258-5437-0 (Hbk)
ISBN 978-3-7258-5438-7 (PDF)
https://doi.org/10.3390/books978-3-7258-5438-7

Contents

About the Editors vii

Preface ix

Vagelis Plevris, Manolis Georgioudakis and Mahdi Kioumarsi
Computational Methods in Structural Engineering: Current Advances and Future Perspectives
Reprinted from: *Computation* **2025**, *13*, 224, https://doi.org/10.3390/computation13090224 . . . 1

Sipho G. Thango, Georgios E. Stavroulakis and Georgios A. Drosopoulos
Investigation of the Failure Response of Masonry Walls Subjected to Blast Loading Using Nonlinear Finite Element Analysis
Reprinted from: *Computation* **2023**, *11*, 165, https://doi.org/10.3390/computation11080165 . . . 8

Spyros Damikoukas and Nikos D. Lagaros
The MLDAR Model: Machine Learning-Based Denoising of Structural Response Signals Generated by Ambient Vibration
Reprinted from: *Computation* **2024**, *12*, 31, https://doi.org/10.3390/computation12020031 30

M. Domaneschi, R. Cucuzza, L. Sardone, S. Londoño Lopez, M. Movahedi and G. C. Marano
Numerical Covariance Evaluation for Linear Structures Subject to Non-Stationary Random Inputs
Reprinted from: *Computation* **2024**, *12*, 50, https://doi.org/10.3390/computation12030050 55

Nikolaos Bakas
Taylor Polynomials in a High Arithmetic Precision as Universal Approximators
Reprinted from: *Computation* **2024**, *12*, 53, https://doi.org/10.3390/computation12030053 69

Lazreg Hadji, Vagelis Plevris, Royal Madan and Hassen Ait Atmane
Multi-Directional Functionally Graded Sandwich Plates: Buckling and Free Vibration Analysis with Refined Plate Models under Various Boundary Conditions
Reprinted from: *Computation* **2024**, *12*, 65, https://doi.org/10.3390/computation12040065 84

Elvis Ababu, George Markou and Sarah Skorpen
Using Machine Learning Algorithms to Develop a Predictive Model for Computing the Maximum Deflection of Horizontally Curved Steel I-Beams
Reprinted from: *Computation* **2024**, *12*, 151, https://doi.org/10.3390/computation12080151 . . . 104

Alireza Mahmoudian, Maryam Bypour and Mahdi Kioumarsi
Explainable Boosting Machine Learning for Predicting Bond Strength of FRP Rebars in Ultra High-Performance Concrete
Reprinted from: *Computation* **2024**, *12*, 202, https://doi.org/10.3390/computation12100202 . . . 133

Khader M. Hamdia
Numerical Homogenization Method Applied to Evaluate Effective Converse Flexoelectric Coefficients
Reprinted from: *Computation* **2025**, *13*, 48, https://doi.org/10.3390/computation13020048 168

About the Editors

Manolis Georgioudakis

Manolis Georgioudakis is a Civil Engineer and holds a Ph.D. in Civil Engineering from the National Technical University of Athens (NTUA). His doctoral dissertation was awarded by the Greek Association of Computational Mechanics (GRACM) as the best PhD Thesis in Computational Methods in Applied Sciences and Engineering. His primary research interests lie in the field of computation mechanics, with particular emphasis on structural optimization, finite element analysis, failure mechanics, the stochastic finite element method, and the applications of machine learning techniques in structural engineering. He actively contributes to the scientific community through publications, international conferences, and collaborative research, and he has been involved in numerous research projects funded by the European Union and Greek national organizations.

Vagelis Plevris

Vagelis Plevris is an Associate Professor in the Department of Civil and Environmental Engineering at Qatar University, where he has been actively engaged in teaching, research, and graduate supervision since 2021. He serves as Specialty Chief Editor of the section "Computational Methods in Structural Engineering" in the journal *Frontiers in Built Environment* and holds editorial positions in several other international journals. His recent research emphasizes the integration of advanced computational methods, optimization, and artificial intelligence in structural engineering, with applications in structural health monitoring, earthquake engineering, sustainable design, topology optimization of civil structures, and the development of explainable AI tools for engineering. He has authored numerous scientific articles and actively contributes to international conferences and collaborative research projects.

Mahdi Kioumarsi

Mahdi Kioumarsi is a Professor in the Department of the Built Environment at Oslo Metropolitan University (OsloMet). He earned his Ph.D. in Structural Engineering from the Norwegian University of Science and Technology (NTNU). With a strong background in structural engineering, Dr. Kioumarsi has established himself as a leading figure in sustainable construction. His academic career spans a broad spectrum of research, ranging from the analysis of large-scale structures to in-depth material investigations, with a particular focus on concrete. Integrating advanced modeling and experimental approaches is central to his research methodology. Over the years, Dr. Kioumarsi has been key in coordinating and contributing to numerous regional, national, and international research projects, collaborating closely with academic institutions and industry partners.

Preface

The Reprint "Computational Methods in Structural Engineering" presents a focused collection of contributions that reflect the evolving role of computation in structural analysis, design, and assessment. The subject of this Reprint is the integration of advanced numerical techniques, optimization strategies, and artificial intelligence into the practice of structural engineering. Its scope extends from traditional finite element modeling and stochastic simulations to innovative data-driven approaches for health monitoring and predictive performance evaluation.

The aim of this Reprint is to highlight recent progress in the field, showcase diverse applications, and provide a resource that supports both academic research and professional practice. The motivation behind this effort is the recognition that modern engineering challenges, ranging from extreme loading events to sustainable design, cannot be addressed without powerful computational tools. By compiling high-quality studies from international researchers, this Reprint seeks to offer a consolidated view of current trends and to stimulate further exploration in areas where computation and engineering intersect.

This Reprint is addressed to scholars, engineers, and graduate students who are engaged in advancing the theory and application of structural engineering. It is also intended for practitioners who wish to gain deeper insight into computational strategies that increasingly shape engineering decision-making. We hope that the contributions gathered here will inspire further innovation and collaboration, while also serving as a valuable reference for those committed to the advancement of structural engineering through computation.

Manolis Georgioudakis, Vagelis Plevris, and Mahdi Kioumarsi
Guest Editors

computation

MDPI

Editorial

Computational Methods in Structural Engineering: Current Advances and Future Perspectives

Vagelis Plevris [1,*], Manolis Georgioudakis [2] and Mahdi Kioumarsi [3]

1 Department of Civil and Environmental Engineering, College of Engineering, Qatar University, Doha P.O. Box 2713, Qatar
2 Institute of Structural Analysis & Antiseismic Research, School of Civil Engineering, National Technical University of Athens, Zografou Campus, 15780 Athens, GR, Greece; geoem@mail.ntua.gr
3 Department of Built Environment, OsloMet—Oslo Metropolitan University, 0166 Oslo, Norway; mahdi.kioumarsi@oslomet.no
* Correspondence: vplevris@qu.edu.qa

Abstract

This brief editorial introduces the Special Issue "Computational Methods in Structural Engineering". This Special Issue brings together recent advances in computational approaches—including finite element modeling, machine learning applications, stochastic analysis, and high-precision numerical methods— highlighting their increasing influence on the analysis, design, and assessment of modern structural systems. The published contributions cover topics such as the nonlinear finite element method (FEM) for structural response under extreme loading, advanced plate and composite modeling, explainable AI for material characterization, machine learning for predictive performance modeling, data-driven signal processing for structural health monitoring, and stochastic analysis of dynamic inputs. Through this collection of studies, this Special Issue underscores both the opportunities and the challenges of applying advanced computational methods to enhance the resilience, efficiency, and understanding of structural engineering systems.

Keywords: structural engineering; computational methods; finite element method; machine learning; explainable AI; structural health monitoring

1. Introduction

Structural engineering has undergone a profound transformation with the rapid advancement and widespread availability of computers. In the past, engineers relied on hand calculations and simplified models, but modern computational methods have now become indispensable in structural analysis and design [1]. The finite element method (FEM), in particular, emerged as a cornerstone of this revolution, providing the "computational workhorse" for simulating complex structures and physics [2]. Introduced in the mid-20th century, the FEM is regarded as one of the most significant engineering advances of the last century, fundamentally changing how engineers model and design structures [3,4]. By discretizing structures into elements, the FEM enables rigorous analysis of stresses, deformations, and failure mechanisms in everything from bridges and high-rise buildings to aerospace and offshore structures. Its development also gave rise to the field of computational mechanics, integrating physics, numerical methods, and computer science in engineering [2]. The result has been a paradigm shift: many structural engineering problems once intractable or oversimplified can now be tackled with high fidelity, leading to safer and more efficient designs in practice [1].

 https://doi.org/10.3390/computation13090224

Equally transformative has been the rise of computational optimization techniques in structural engineering [5–7]. Optimization algorithms allow engineers to automatically search for designs that minimize the weight, cost, or environmental impact, while satisfying the safety constraints. Early applications involved sizing or shape optimization, but the field truly accelerated with topology optimization—a method that optimizes the material layout within a design space. Topology optimization techniques, pioneered in the 1980s and 1990s, have enabled the creation of lightweight yet strong structural forms by removing unnecessary material [8]. Bendsøe and Sigmund's foundational work introduced general methods to optimize structural layouts, achieving minimum-weight designs without compromising performance [8,9]. Over the past two decades, numerous advances have made such techniques more powerful and practical [5,10]. Comprehensive surveys have shown that modern topology and shape optimization methods—ranging from gradient-based approaches to evolutionary algorithms—are now successfully applied to real-world structures across domains such as aerospace engineering and civil infrastructure [5,6]. These computational design tools improve the efficiency and material usage in structures and also open up innovative architectural forms that would be impossible to realize via manual trial and error [11].

In parallel, performance-based design has emerged as a modern paradigm, particularly in earthquake [12,13] and fire engineering [14], in which structures are iteratively engineered to meet explicit performance targets (e.g., life safety or immediate occupancy) under extreme loads. Such design approaches rely heavily on nonlinear finite element simulations and probabilistic analyses, which are only feasible thanks to today's computational power [14,15]. Advanced numerical models also tackle complex materials and geometries. For instance, refined finite element formulations allow the analysis of composite structures and unconventional shapes that exhibit behaviors beyond the scope of classical beam or plate theory. From multi-directional functionally graded materials to long-span free-form shells, computational methods provide the means to predict structural responses with confidence where analytical solutions are unavailable [16].

In recent years, the rapid development of machine learning (ML) and artificial intelligence has further expanded the computational toolkit of structural engineers. ML techniques are now being used as powerful supplements (and sometimes alternatives) to physics-based simulations in various structural applications [17,18]. In the last decade, there has been a boom in implementing data-driven models for tasks such as structural health monitoring, damage detection, predictive modeling of structural behavior, and even design automation [17,19,20]. Unlike traditional programs written from first principles, ML algorithms can learn complex nonlinear relationships directly from data—a capability especially useful for problems where accurate analytical modeling is difficult [21]. For example, researchers have trained neural networks and ensemble methods to predict structural responses (deflections, stresses, failures) under loads, bypassing more time-consuming finite element analyses in specific scenarios [17]. Likewise, ML-driven models have achieved remarkable accuracy in estimating material properties (e.g., concrete strength, bond capacity of novel materials) by mining large experimental datasets [22,23]. These approaches can handle high-dimensional data (e.g., vibration signals, monitoring sensor streams) and filter out noise or detect patterns that elude conventional methods [17,24]. While data-driven models do not replace the need for fundamental mechanics, they complement traditional simulations—for instance, hybrid approaches use ML as fast surrogates for costly computations or to optimize structural systems in real time [25]. The net effect is that structural engineering is becoming more predictive and adaptive [17]. Of course, challenges remain in ensuring the reliability, interpretability, and adequate training of these models, but ongoing research is quickly advancing the integration of AI into structural engineering workflows [24].

Structural health monitoring (SHM) deserves special mention as one of the most active areas of computational innovation in structural engineering. Modern SHM frameworks combine vibration-based testing, dense sensor networks, and advanced data analytics to assess the integrity and performance of structures in real time [26]. With the integration of machine learning, SHM systems can process high-dimensional and noisy monitoring data, enabling more reliable detection of anomalies and early identification of damage [21,22]. These advances support the development of predictive maintenance and early-warning systems that enhance the safety, resilience, and lifecycle management of critical infrastructure [27]. The inclusion of SHM-related contributions in this Special Issue reflects its growing significance in bridging advanced computational methods with practical asset management.

Another frontier in computational structural engineering lies in the advancement of stochastic methods and high-precision numerical techniques. Real-world structures are inherently affected by uncertainties in loads, materials, and environmental conditions, which makes uncertainty quantification and probabilistic modeling essential [28]. Modern approaches such as reliability-based design [29], stochastic dynamic analysis [30], and covariance evaluation provide rigorous tools to capture the variability and assess the structural safety under realistic time-varying conditions. In parallel, progress in high-precision numerical methods has expanded the accuracy and stability of classical formulations, allowing engineers to tackle nonlinear, multi-scale, and multi-physics problems with greater fidelity [31,32]. Together, these developments highlight how computational innovation—both in handling uncertainty and in refining numerical precision—continues to broaden the scope and reliability of structural engineering analysis.

Motivated by these advancements, this Special Issue, "Computational Methods in Structural Engineering", was conceived to showcase recent developments and applications at the forefront of the field. The call for papers solicited contributions across a wide range of topics—including finite element analysis, optimization techniques, dynamic simulation, novel numerical methods for materials and forms, and machine learning applications—reflecting the broad impact of computation in modern structural engineering. The response to the call has been satisfactory, leading to eight high-quality peer-reviewed papers that collectively showcase some of the most relevant research trends. These contributions span from high-fidelity nonlinear finite element modeling of structural failure under extreme loads to data-driven approaches for processing structural response signals and predicting structural performance, to advanced analytical and numerical techniques for innovative materials and systems. In the following sections, we summarize each contribution to this Special Issue, emphasizing its context and key findings within the broader evolution of computational structural engineering.

2. Contributions

Thango et al. (Contribution 1) investigate the failure response of masonry walls subjected to blast loading using nonlinear finite element analysis (FEM). Masonry structures, widely used in many regions, are highly vulnerable to extreme dynamic loads such as explosions. The authors employ advanced nonlinear FEM models to simulate the blast-induced behavior of masonry walls, capturing complex phenomena including the material nonlinearity, cracking, and progressive failure. The study provides detailed insights into how different parameters, such as loading intensity and boundary conditions, influence wall performance under blast effects. The findings highlight the capability of computational modeling to replicate failure mechanisms that are otherwise difficult to observe experimentally due to safety and cost constraints. This contribution emphasizes the importance of nonlinear FEM in understanding structural resilience under extreme events, offering valuable guidance for protective design and retrofitting strategies.

Damikoukas and Lagaros (Contribution 2) propose the MLDAR model, a machine learning-based framework for denoising structural response signals generated by ambient vibration testing. Structural health monitoring (SHM) often relies on ambient vibration data, but such signals are prone to noise, which can hinder accurate modal identification and damage detection. Traditional filtering methods, such as Fourier or wavelet transforms, have been widely used to address this issue; however, modern machine learning techniques increasingly outperform them. In this work, the authors develop a denoising method that integrates ML algorithms to separate meaningful structural response components from background noise. Their results show that the proposed MLDAR model significantly improves the signal clarity compared to conventional filtering approaches, enabling more reliable extraction of modal parameters. By enhancing the quality of vibration-based data, this study contributes to more robust SHM applications and demonstrates the effectiveness of data-driven approaches in complementing traditional signal processing. The work underscores the growing role of machine learning in improving data quality for structural assessment and monitoring.

Domaneschi et al. (Contribution 3) address the challenge of evaluating covariance in linear structural systems subjected to non-stationary random inputs. Many real-world structural systems are influenced by time-varying dynamic loads, such as earthquakes, wind gusts, or traffic, where classical stationary assumptions no longer hold. The authors propose a numerical method for efficiently computing covariance responses under such non-stationary excitations, providing a valuable tool for uncertainty quantification and reliability analysis. The framework allows for the accurate characterization of structural response statistics without resorting to prohibitively large Monte Carlo simulations. Applications of the method demonstrate its potential for analyzing the safety and serviceability of engineering structures under realistic loading conditions. This contribution reinforces the importance of advanced stochastic analysis in structural engineering, offering a computationally efficient approach to assess performance under complex and uncertain environments.

Bakas (Contribution 4) investigates using Taylor polynomials under high arithmetic precision as universal function approximators. The study revisits the classical concept of Taylor series expansion and explores its potential in modern computational contexts, particularly where high numerical precision is required. By employing advanced numerical techniques, the work demonstrates that Taylor polynomials can serve as reliable universal approximators, effectively capturing complex nonlinear behaviors with enhanced stability and accuracy. This approach offers a promising alternative to more computationally intensive approximation schemes, such as spectral or mesh-based methods, particularly in cases where analytical solutions are intractable or conventional numerical techniques may suffer from precision loss. This contribution highlights how fundamental mathematical tools, combined with high-precision computing, can provide robust and efficient approximation strategies for diverse engineering and scientific applications.

Hadji et al. (Contribution 5) focus on the buckling and free vibration behavior of multi-directional functionally graded (FG) sandwich plates, analyzed using refined plate theories under various boundary conditions. Functionally graded sandwich plates, which combine the advantages of FG materials and sandwich structures, are increasingly studied for their superior mechanical performance in aerospace, civil, and mechanical applications. The authors employ advanced refined plate models to capture the effect of material gradation in multiple directions, allowing for a more realistic representation of the stiffness and mass distribution. Their analysis explores how different boundary conditions influence both the critical buckling loads and natural frequencies of these innovative structures. The findings provide valuable insights into the stability and dynamic characteristics of FG sandwich plates, emphasizing the importance of considering material gradation patterns in both directions. This research extends the applicability of computational modeling to advanced composite structures and contributes to the broader field of multi-scale modeling of innovative structural systems.

Ababu et al. (Contribution 6) present a study on the use of machine learning algorithms to develop predictive models for estimating the maximum deflection of horizontally curved steel I-beams. Curved I-beams are widely used in bridges and other infrastructures where alignment requires nonlinear geometry. Still, their structural response under loading is significantly more complex than straight beams. Accurate traditional analytical or numerical methods can be computationally intensive and require specialized expertise. In this work, the authors apply supervised ML algorithms to experimental and simulation datasets, creating models that accurately predict the maximum deflections. The study highlights the efficiency of data-driven approaches in capturing the nonlinear relationships inherent in curved beam mechanics. Beyond predictive accuracy, the results demonstrate the potential of ML-based surrogates to reduce the computational costs in design and assessment processes. This contribution illustrates how modern data science can complement classical structural mechanics, offering engineers faster and more accessible predictive tools for complex structural elements.

Mahmoudian et al. (Contribution 7) investigate the use of explainable boosting machine (EBM) models to predict the bond strength of fiber-reinforced polymer (FRP) rebars embedded in ultra-high-performance concrete (UHPC). The bond strength between FRP reinforcement and concrete is a key factor influencing the structural performance, durability, and service life, especially given the increasing use of FRP bars as corrosion-resistant alternatives to steel. In this study, the authors apply EBM, a state-of-the-art machine learning approach that combines predictive power with interpretability, allowing the identification of the most influential parameters affecting bond strength. The model demonstrates excellent predictive accuracy compared to traditional regression methods and provides transparent insights into the relative importance of input variables such as the concrete compressive strength, rebar surface characteristics, and embedment length. By offering accurate predictions and clear explanations of the underlying data-driven relationships, the work advances the integration of trustworthy AI tools in structural engineering practice. This contribution is particularly valuable for guiding design decisions and developing future design codes involving FRP-reinforced UHPC structures.

Hamdia (Contribution 8) presents a study on the application of a numerical homogenization method to evaluate the effective converse flexoelectric coefficients of composite materials. Flexoelectricity, a size-dependent electromechanical coupling phenomenon, is particularly significant in small-scale materials, where strain gradients can induce electric polarization. This work employs a computational framework to bridge the gap between microscale material behavior and effective macroscale properties, enabling accurate prediction of converse flexoelectric coefficients. The study emphasizes the versatility of numerical homogenization techniques in handling complex material microstructures, which are often challenging to address analytically. The results highlight the capacity of this approach to capture essential electromechanical interactions, providing insights into the design and optimization of advanced multifunctional materials. The findings expand the understanding of flexoelectric phenomena and open opportunities for developing innovative smart structures and devices where electromechanical coupling plays a central role. This contribution underscores the growing role of computational homogenization in linking material science with structural engineering applications.

Author Contributions: All authors contributed to every part of the research described in this paper. All authors have read and agreed to the published version of the manuscript.

Funding: This research received no external funding.

Acknowledgments: This Special Issue has been made possible by the contributions of several authors, reviewers, and editorial team members. Their efforts are greatly appreciated, and the Guest Editors would like to express their sincere thanks to all of them.

Conflicts of Interest: The authors declare no conflicts of interest.

List of Contributions:

1. Thango, S.G.; Stavroulakis, G.E.; Drosopoulos, G.A. Investigation of the Failure Response of Masonry Walls Subjected to Blast Loading Using Nonlinear Finite Element Analysis. *Computation* **2023**, *11*, 165. https://doi.org/10.3390/computation11080165.
2. Damikoukas, S.; Lagaros, N.D. The MLDAR Model: Machine Learning-Based Denoising of Structural Response Signals Generated by Ambient Vibration. *Computation* **2024**, *12*, 31. https://doi.org/10.3390/computation12020031.
3. Domaneschi, M.; Cucuzza, R.; Sardone, A.; Lopez, S.L.; Movahedi, M.; Marano, G.C. Numerical Covariance Evaluation for Linear Structures Subject to Non-Stationary Random Inputs. *Computation* **2024**, *12*, 50. https://doi.org/10.3390/computation12030050.
4. Bakas, N. Taylor Polynomials in a High Arithmetic Precision as Universal Approximators. *Computation* **2024**, *12*, 53. https://doi.org/10.3390/computation12030053.
5. Hadji, L.; Plevris, V.; Madan, R.; Ait Atmane, H. Multi-Directional Functionally Graded Sandwich Plates: Buckling and Free Vibration Analysis with Refined Plate Models under Various Boundary Conditions. *Computation* **2024**, *12*, 65. https://doi.org/10.3390/computation12040065.
6. Ababu; Markou, G.; Skorpen, S. Using Machine Learning Algorithms to Develop a Predictive Model for Computing the Maximum Deflection of Horizontally Curved Steel I-Beams. *Computation* **2024**, *12*, 151. https://doi.org/10.3390/computation12080151.
7. Mahmoudian, A.; Bypour, M.; Kioumarsi, M. Explainable Boosting Machine Learning for Predicting Bond Strength of FRP Rebars in Ultra High-Performance Concrete. *Computation* **2024**, *12*, 202. https://doi.org/10.3390/computation12100202.
8. Hamdia, K.M. Numerical Homogenization Method Applied to Evaluate Effective Converse Flexoelectric Coefficients. *Computation* **2025**, *13*, 48. https://doi.org/10.3390/computation13020048.

References

1. Owen, D.R.J.; Feng, Y.T. Fifty years of finite elements—A solid mechanics perspective. *Theor. Appl. Mech. Lett.* **2012**, *2*, 051001. [CrossRef]
2. Zienkiewicz, O.C.; Taylor, R.L.; Zhu, J.Z. *The Finite Element Method: Its Basis and Fundamentals*; Elsevier: Amsterdam, The Netherlands, 2013. [CrossRef]
3. Clough, R.W. The finite element method after twenty-five years: A personal view. *Comput. Struct.* **1980**, *12*, 361–370. [CrossRef]
4. Belytschko, T.; Liu, W.K.; Moran, B.; Elkhodary, K. *Nonlinear Finite Elements for Continua and Structures*, 2nd ed.; Wiley: Hoboken, NJ, USA, 2014.
5. Deaton, J.D.; Grandhi, R.V. A survey of structural and multidisciplinary continuum topology optimization: Post 2000. *Struct. Multidisc. Optim.* **2014**, *49*, 1–38. [CrossRef]
6. Rozvany, G.I.N. A critical review of established methods of structural topology optimization. *Struct. Multidisc. Optim.* **2009**, *37*, 217–237. [CrossRef]
7. Papadrakakis, M.; Lagaros, N.D.; Tsompanakis, Y.; Plevris, V. Large Scale Structural Optimization: Computational Methods and Optimization Algorithms. *Arch. Comput. Methods Eng.* **2001**, *8*, 239–301. [CrossRef]
8. Bendsøe, M.P. Optimal shape design as a material distribution problem. *Struct. Optim.* **1989**, *1*, 193–202. [CrossRef]
9. Bendsøe, M.P.; Sigmund, O. *Topology Optimization: Theory, Methods, and Applications*; Springer: Berlin/Heidelberg, Germany, 2004. [CrossRef]

10. Chamatidis, I.; Stoumpos, M.; Kazakis, G.; Kallioras, N.A.; Triantafyllou, S.; Plevris, V.; Lagaros, N.D. Overview on Machine Learning Assisted Topology Optimization Methodologies. In *Machine Learning in Modeling and Simulation: Methods and Applications*; Rabczuk, T., Bathe, K.-J., Eds.; Springer International Publishing: Cham, Switzerland, 2023; pp. 373–394. [CrossRef]
11. Plevris, V.; Almutairi, A.; Jiménez Rios, A. Advancing Sustainability Through Structural Optimization: Innovations in Material Efficiency and Environmental Impact Reduction. In Proceedings of the 1st International Conference on Net-Zero Built Environment: Innovations in Materials, Structures, and Management Practices (Net-Zero Future 2024), Oslo, Norway, 19–21 June 2024; pp. 1611–1623.
12. Chandler, A.M.; Lam, N.T.K. Performance-based design in earthquake engineering: A multi-disciplinary review. *Eng. Struct.* **2001**, *23*, 1525–1543. [CrossRef]
13. Ghobarah, A. Performance-based design in earthquake engineering: State of development. *Eng. Struct.* **2001**, *23*, 878–884. [CrossRef]
14. Franchini, A.; Galasso, C.; Torero, J.L. Probabilistic Performance-based Fire Design of Structures: A Hazard-Centred and Consequence-Oriented Perspective. *Fire Technol.* **2024**, *60*, 2845–2873. [CrossRef]
15. Hu, S.; Guo, T.; Alam, M.S.; Koetaka, Y.; Ghafoori, E.; Karavasilis, T.L. Machine learning in earthquake engineering: A review on recent progress and future trends in seismic performance evaluation and design. *Eng. Struct.* **2025**, *340*, 120721. [CrossRef]
16. Ghatage, P.S.; Kar, V.R.; Sudhagar, P.E. On the numerical modelling and analysis of multi-directional functionally graded composite structures: A review. *Compos. Struct.* **2020**, *236*, 111837. [CrossRef]
17. Salehi, H.; Burgueño, R. Emerging artificial intelligence methods in structural engineering. *Eng. Struct.* **2018**, *171*, 170–189. [CrossRef]
18. Bypour, M.; Yekrangnia, M.; Kioumarsi, M. Machine learning-driven optimization for predicting compressive strength in fly ash geopolymer concrete. *Clean. Eng. Technol.* **2025**, *25*, 100899. [CrossRef]
19. Thai, H.-T. Machine learning for structural engineering: A state-of-the-art review. *Structures* **2022**, *38*, 448–491. [CrossRef]
20. Plevris, V.; Ahmad, A.; Lagaros, N.D. (Eds.) *Artificial Intelligence and Machine Learning Techniques for Civil Engineering*; IGI Global: Pennsylvania, PA, USA, 2023. [CrossRef]
21. Cha, Y.-J.; Ali, R.; Lewis, J.; Büyüköztürk, O. Deep learning-based structural health monitoring. *Autom. Constr.* **2024**, *161*, 105328. [CrossRef]
22. Azimi, M.; Eslamlou, A.D.; Pekcan, G. Data-Driven Structural Health Monitoring and Damage Detection through Deep Learning: State-of-the-Art Review. *Sensors* **2020**, *20*, 2778. [CrossRef]
23. Mahmoudian, A.; Bypour, M.; Kioumarsi, M. Explainable Boosting Machine Learning for Predicting Bond Strength of FRP Rebars in Ultra High-Performance Concrete. *Computation* **2024**, *12*, 202. [CrossRef]
24. Plevris, V. AI-Driven Innovations in Earthquake Risk Mitigation: A Future-Focused Perspective. *Geosciences* **2024**, *14*, 244. [CrossRef]
25. Solorzano, G.; Plevris, V. ANN-based surrogate model for predicting the lateral load capacity of RC shear walls. In Proceedings of the 8th European Congress on Computational Methods in Applied Sciences and Engineering (ECCOMAS 2022), Oslo, Norway, 5–9 June 2022.
26. Inaudi, D. Structural health monitoring of bridges: General issues and applications. In *Structural Health Monitoring of Civil Infrastructure Systems*; Karbhari, V.M., Ansari, F., Eds.; Woodhead Publishing: Cambridge, UK, 2009; pp. 339–370. [CrossRef]
27. Plevris, V.; Papazafeiropoulos, G. AI in Structural Health Monitoring for Infrastructure Maintenance and Safety. *Infrastructures* **2024**, *9*, 225. [CrossRef]
28. Bianchini, I.; Argiento, R.; Auricchio, F.; Lanzarone, E. Efficient uncertainty quantification in stochastic finite element analysis based on functional principal components. *Comput. Mech.* **2015**, *56*, 533–549. [CrossRef]
29. Hu, W.; Cheng, S.; Yan, J.; Cheng, J.; Peng, X.; Cho, H.; Lee, I. Reliability-based design optimization: A state-of-the-art review of its methodologies, applications, and challenges. *Struct. Multidisc. Optim.* **2024**, *67*, 168. [CrossRef]
30. Yang, D.; Chen, G. Stochastic Dynamic Analysis of Large-Scale Nonlinear Structures. In *Advances in Applied Nonlinear Dynamics, Vibration and Control-2021*; Springer: Singapore, 2022; pp. 845–859. [CrossRef]
31. Barchiesi, E. Multi-scale and multi-physics: Towards next-generation engineering materials. *Contin. Mech. Thermodyn.* **2020**, *32*, 541–554. [CrossRef]
32. Bayat, M.; Dong, W.; Thorborg, J.; To, A.C.; Hattel, J.H. A review of multi-scale and multi-physics simulations of metal additive manufacturing processes with focus on modeling strategies. *Addit. Manuf.* **2021**, *47*, 102278. [CrossRef]

computation

MDPI

Article

Investigation of the Failure Response of Masonry Walls Subjected to Blast Loading Using Nonlinear Finite Element Analysis

Sipho G. Thango [1], Georgios E. Stavroulakis [2,*] and Georgios A. Drosopoulos [1,3]

[1] Discipline of Civil Engineering, University of Kwazulu-Natal, Durban 4041, South Africa; 209523102@stu.ukzn.ac.za (S.G.T.); gdrosopoulos@uclan.ac.uk (G.A.D.)
[2] Department of Production Engineering and Management, Technical University of Crete, 73100 Chania, Greece
[3] Discipline of Civil Engineering, University of Central Lancashire, Preston PR1 2HE, UK
* Correspondence: gestavroulakis@tuc.gr; Tel.: +30-2821037418

Abstract: A numerical investigation of masonry walls subjected to blast loads is presented in this article. A non-linear finite element model is proposed to describe the structural response of the walls. A unilateral contact–friction law is used in the interfaces of the masonry blocks to provide the discrete failure between the blocks. A continuum damage plasticity model is also used to account for the compressive and tensile failure of the blocks. The main goal of this article is to investigate the different collapse mechanisms that arise as an effect of the blast load parameters and the static load of the wall. Parametric studies are conducted to evaluate the effect of the blast source–wall (standoff) distance and the blast weight on the structural response of the system. It is shown that the traditional in-plane diagonal cracking failure mode may still dominate when a blast action is present, depending on the considered standoff distance and the blast weight when in-plane static loading is also applied to the wall. It is also highlighted that the presence of an opening in the wall may significantly reduce the effect of the blasting action.

Keywords: masonry; collapse mechanism; blast actions; unilateral contact; dynamic analysis; finite element analysis

1. Introduction

Over the years, increasing research efforts have been developed focusing on analyzing the structural behavior of masonry walls. Masonry is defined as a set of stone units that are connected using mortar joints that are organized to form a regular pattern [1]. Masonry is commonly used in monuments, masonry arches, and also in low-cost houses. For these structural systems, the low tensile resistance of masonry or mortar interfaces may lead to a compromised response when in- and out-of-plane lateral forces reach high values.

Among several loading conditions, ongoing research aims to investigate the impact of blasting forces on masonry structures. In particular, research on this type of loading focuses on mining activities using blasting operations, which comprise the first phase of the production cycle in most of the mining processes. Blasting is used to fragment the rock overlying the coal seams in most mines. When the explosives are detonated, most of the energy is consumed in rock fragmentation [2]. According to [3], energy not used to break rock radiates out from the blast site in the form of ground vibrations and air blasts. Additionally, when explosives are ignited in rock, a shock wave is produced that breaks the rock and then a force in the form of gas pressure is formed [4]. An explosion or blast activity is defined as the release of a significant amount of energy that takes place in a short time period.

Computer advancement in the past decades has enabled researchers to model masonry with its complexities using finite element analysis. The finite element (FE) method is one of the advanced numerical techniques that is commonly applied to analyze complex structural engineering problems. Research work presented in [5–9] and others indicate that by using the FE method, the failure modes that occur in masonry due to blast loading can be successfully analyzed.

According to [10], the collapse modes of masonry walls that are exposed to blast actions may include flexural failure, direct shear failure, and flexural–shear failure. Collapse modes are further discussed in this section and elaborated in the analysis section of this article. It is noted that these collapse modes were used for the validation of the proposed numerical model.

D'Altri et al. [11] considered a masonry wall with dimensions of 1190 mm × 795 mm with a brick size of 112 mm × 53 mm × 36 mm. The boundary conditions were taken as fixed on all four sizes of the wall. The wall was loaded with a 20 KN/m^2 out-of-plane load. Their research aimed to assess the effectiveness of the micro-modeling approach and assess the out-of-plane response of the masonry walls. As defined by Lourenço [12], micro-modeling is where "masonry units and mortar joints are represented by continuum elements, where the unit-mortar interface is represented by a discontinuous constitutive description". A quasi-static (transient dynamic) procedure was used for the numerical study. Furthermore, the brick–mortar bond failures were accounted for using brick–mortar nonlinear cohesive interfaces. The failure pattern in the wall indicated that the maximum displacement often occurs at the center of the wall.

The discrete element method was used in [13] to investigate the behavior of masonry structures under blast actions. A 2400 mm × 2400 mm wall, fixed on all sides, was simulated, and typical modes of failure, including out-of-plane failure, were observed. Furthermore, the study depicted the complete failure of the wall under a load of 810 kg TNT explosive weight at a standoff distance of 37 m. According to Masi et al. [13], the geometry of the blocks and the interfaces may be directly modeled using the discrete element method. Their study was conducted using 3DEC software and the empirical model CONWEP to simulate the blast action. They used a soft-contact technique to simulate joint interactions between adjacent blocks. It is worth mentioning that the magnitude of the wall failure is dependent on various factors such as standoff distance, wall dimensions/properties, and boundary conditions.

Hao [5] conducted a numerical analysis of a 2880 mm × 2820 mm masonry wall subject to blast load corresponding to a TNT explosive weight W = 2000 kg using AUTODYN software. In that study, the four sides of the wall were modeled as fixed, with a mortar layer between the fixed boundary and the masonry units of the wall, which, in turn, was assigned homogenized material properties. It was shown that for higher explosive weight and shorter standoff distances, the wall would collapse, and the center portion of the wall failed out-of-plane as one brick flew out as a single piece. The wall was also observed to be damaged near the boundary.

Shamim et al. [14] conducted a numerical study investigating the effect of a blast on a 3000 mm × 3000 mm × 230 mm masonry wall, which had a reinforced concrete frame of 230 mm × 235 mm cross-section dimensions. In their macro-approach, masonry units, mortar joints, and the brick–mortar interface were modeled as a single material. They investigated the effect of 100 kg TNT explosive weight over 20 m, 30 m, and 40 m distances from the wall. Furthermore, their study considered a wall without an opening as well as a wall with a window opening at its center. The boundary conditions were defined such that the top of the wall was restrained in the direction parallel to the blast, simulating the restrain obtained from a slab due to its high in-plane stiffness. The results for the wall without the window showed that the peak values of displacements are found at mid-span. The peak displacement values were equal to 267.8 mm, 95.1 mm, and 59.9 mm for the three mentioned standoff distances between the blast source and the wall, respectively. For the wall with the window, the values of peak displacement at the top of the opening when out-of-plane failure arose were equal to 353.6 mm, 121.9 mm, and 73.2 mm, respectively. Overall, they observed that peak values on the wall with the window were higher than those of the wall without the window; however, the standoff distances were not the same when the opening was considered.

In a similar investigation presented in [15], it was shown that a wall subjected to blast actions developed the highest displacement in the midsection of the masonry infill panel, while the reinforced concrete frame remained undamaged. It was shown that when the blasting source was close to the wall, the masonry panel collapsed completely, depicting displacements greater than the thickness of the wall (>230 mm). Their study also looked at the effect of changing blast load sizes, considering a TNT equivalent weight of charge equal to W1 = 25 kg, W2 = 50 kg, W3 = 75 kg, and W4 = 100 kg for a constant standoff distance of 20 m. It was observed that peak displacement increases with increasing weight of charge (at constant standoff distance of 20 m) and decreases with increasing distance.

In [16], a numerical study was conducted on a masonry wall with dimensions of 1700 mm × 1550 mm × 100 mm. The model was constructed with 23 courses of solid clay bricks and analyzed using a simplified micro-modeling approach within finite element analysis. The simulation was implemented in steps, involving vertical displacements and cyclic out-of-plane actions. The failure mode was due to the formation of diagonal cracks caused by in-plane loading. As derived from the mentioned literature, the type of failure modes of masonry walls under in-plane and blasting, out-of-plane loading, are influenced by various characteristics—such as the load application, geometry, boundary conditions, and the quality of materials.

Some recent efforts aim at investigating the response of different types of reinforced masonry walls under blast actions. In [17], a masonry wall connected with two transverse walls, one at each end, was numerically tested using the micro-modeling technique within non-linear finite element analysis. The work proposed numerical models to reinforce the wall using CFRP wrapping and a steel angle-strip system. In [18], the behavior of unreinforced masonry walls with CFRP wrapping and mild welded steel wire mesh, under blast with low standoff distance, was investigated using non-linear finite element analysis. In [19], a fragility analysis of masonry walls was proposed, illustrating the vulnerability of the structures against blast load, focusing on different types of unreinforced masonry walls and reinforced walls, using finite element analysis. In [20], for masonry walls made of autoclaved aerated concrete and polymer-reinforced concrete that are subjected to heavy TNT explosive loads, both experimental and numerical testing were provided. In the numerical models, non-linear finite element analysis was used with cohesive zone models to depict damage to the wall.

Based on this short review of recent results, it seems that there is still space for more research investigating the collapse modes of masonry walls under blast actions. In particular, one of the goals of this article, which also highlights its innovative points, is to provide further insight into the way in-plane failure modes, such as diagonal cracking and out-of-plane damage, may appear in masonry walls subjected to blast actions. From another point of view, this article proposes a modeling technique using non-linear constitutive descriptions, incorporating opening-sliding failure modes adopting contact mechanics, as well as compressive/tensile damage, using continuum damage laws, all within finite element analysis. The proposed models can be implemented in commercial software.

Within the given framework, a numerical investigation of the mechanical response of masonry walls under blast actions, with and without openings, is presented. Non-linear finite element models are proposed to simulate all the joints between masonry units by introducing unilateral contact–friction interfaces. For the simulation of the blast action, an empirical model is used, and explicit dynamic analysis is adopted implementing this loading type. Various loading cases are tested, resulting in different failure modes.

In Section 2 of this article, failure modes of masonry walls are provided and modeling approaches that can be used to capture these modes are briefly discussed. In Section 3, all the details of the numerical model that is proposed in this article are presented. Among others, the details of the blast load simulation, the material constitutive description, and the geometry of the walls are given in this section. In Section 4, a validation of the proposed model is conducted using a comparison of some results with published output. In Section 5, results and discussions derived from the suggested approach are provided, and in Section 6, the conclusions of this investigation are presented.

2. Failure Modes and Modeling Approaches of Masonry Walls

In this section, a preliminary discussion of failure modes for masonry walls that occur under various loading scenarios is provided. Both in-plane and out-of-plane damage patterns are recognized. In the next sections, it will be shown that some of these modes arise also under blast actions, depending on the load combination.

In addition, general concepts elaborating modeling approaches, which are used to capture the mentioned failure modes, are discussed. Within this framework, the proposed model will be identified.

2.1. In-Plane Response of Masonry Walls

Three types of failure modes of masonry walls under static loading are discussed below, and these are sliding shear, flexural failure, and diagonal shear. These failure modes are illustrated in Figure 1.

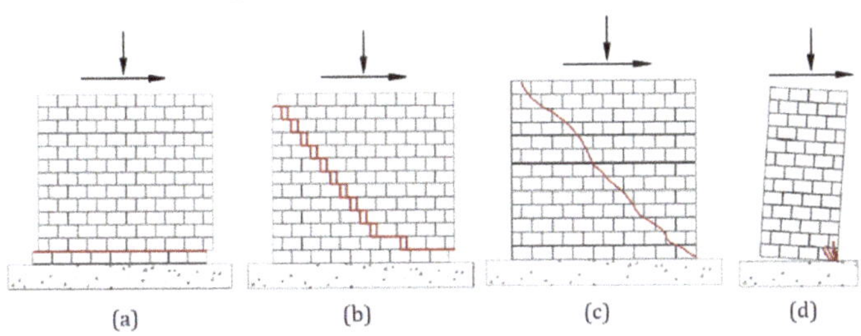

Figure 1. Typical Failure modes of masonry walls subjected to a vertical load and a horizontal (shear) load. These two loads will result in (**a**) sliding shear failure, (**b**) shear failure (staircase-shaped cracks), (**c**) diagonal shear, and (**d**) crushing, which is mainly compressive cracks [21].

According to [21], the in-plane failure of a wall is generally shear failure, as can be seen by the diagonal cracking, and this is often ruled by the tensile resistance or capacity of the masonry unit or mortar joints. Additionally, a wall's failure can be observed as the crushing of units under compression. The capacity for large displacement and average energy dissipation is related to unit crushing [22], contrary to failure involving sliding shear, which is more ductile as a higher amount of energy is dissipated. This energy dissipation is more common under seismic/blast loading actions. The above modes of failure have been proven by many researchers to be the common failure modes when horizontal displacement and vertical pressure loading are applied to walls.

2.2. Out-of-Plane Response of Masonry Walls

Subject to blast loadings, the failure mode of masonry walls is often out-of-plane flexural failure. This may be accompanied by a flexural cracking pattern, which consists of horizontal cracks arising at halfway of the wall and stepped diagonal cracks toward all corners of the wall, as can be seen in Figure 2, where a typical out-of-plane response of a masonry wall under blast loading is shown. It is noted that these descriptions of the failure modes will be used later in this article to verify the results that are obtained from the proposed numerical scheme.

Figure 2. Out-of-plane failure pattern of a wall subjected to blast loading.

2.3. General Modeling Approaches for Masonry Structures

The mechanical behavior of masonry buildings has been described using two broad numerical approaches: macro-modeling and micro-modeling [12]. In the macro-modeling method, masonry is analyzed as a uniform material that obtains its average (effective) material properties by a homogenization scheme. Thus, in the macro-modeling technique, masonry units (concrete blocks, stone units) and the mortar joints are modeled as solitary materials using homogenization concepts. According to [23], when using the macro-modeling approach, the detailed failure mechanisms may generally not be reproduced well.

According to micro-modeling, masonry consists of joints connecting individual units, and using appropriate constitutive laws, simulation of such walls is conducted. Due to the different compression or tensile strength of brick vs. mortar, it is worth noting that mortar joints become the weakest link in masonry walls. According to [12], the unit–mortar interface controls the nonlinear response of the joints, and this is one of the most pertinent features of masonry wall behavior. Different modeling techniques used to simulate the response of masonry structures are depicted schematically in Figure 3.

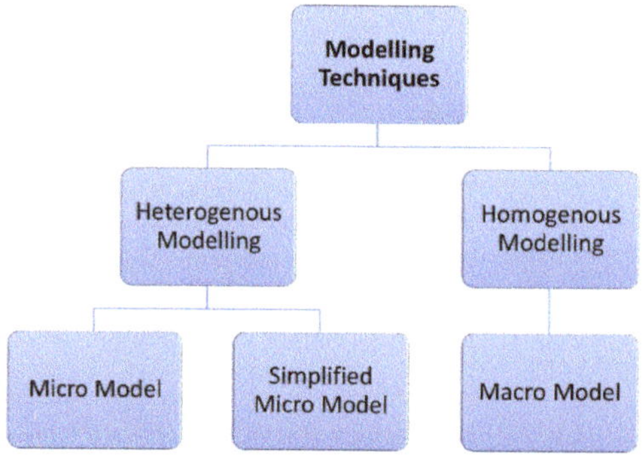

Figure 3. Modeling techniques for masonry walls.

The masonry units are modeled as continuum elements, while the mortar joints are modeled as interface elements in this article. Zero tensile resistance between the joints is introduced, using unilateral contact and friction interfaces. More details about the model used in this article are provided in the following sections.

3. The Numerical Model Proposed in the Present Article

A non-linear finite element model is proposed for this study to simulate the response of masonry walls to blasts. For the evaluation of the failure response that is derived from the masonry unit interfaces, a unilateral contact and friction constitutive description is assigned to these interfaces. Thus, both in-plane and out-of-plane opening and/or sliding between the masonry units can be depicted with the proposed model. Since all the contact conditions between the blocks in the wall are simulated using principles taken from contact mechanics, it can be stated that the micro-modeling approach is used.

The unilateral contact law, provided in Equations (1)–(3) for a single degree of freedom system, is assigned in the interfaces between masonry blocks. Equation (1) is the non-penetration relation, Equation (2) states that only compressive stresses (t^n) can be developed in the interfaces, and Equation (3) is the complementarity relation, stating that either contact takes place ($u - g = 0$, where u is a single degree of freedom and g is an initial gap) or separation in the interface occurs ($t^n = 0$).

$$h = u - g \leq 0 \Rightarrow h \leq 0 \tag{1}$$

$$-t^n \geq 0 \tag{2}$$

$$t^n (u - g) = 0 \tag{3}$$

For the response in the tangential direction of the interfaces, a static version of Coulomb's friction law is considered. Thus, sliding in the interfaces is initiated when the shear stress t^t reaches the critical value τ_{cr}, according to Equation (4):

$$t^t = \tau_{cr} = \pm\mu|t^n| \tag{4}$$

where μ is the friction coefficient and t^n the normal stress (contact pressure) in the interfaces.

To represent the failure response of the masonry units, a continuum concrete damage plasticity model is used. Compressive and tensile failure modes developed at the masonry blocks are then depicted. In the following sections, the details related to the implementation of the blast loading, the material properties, and the dimensions of the walls that are studied in this article are provided.

3.1. Blast Shock Wave Modeling

An explosion loading wave is defined by three parameters, namely, the shape of a wave, the maximum pressure (P_{ro}), and the positive wave duration (t_o), which is the time that pressure reaches zero [24]. Various research efforts have shown that depending on the source of the explosion, the generated waves are divided into shock and pressure waves. In a shock wave, the pressure of gasses from the explosion or blasting is developed by emission from the source of the explosion [24,25]. The pressure increases to the maximum value P_{ro} and decreases to the environmental pressure, as shown in Figure 4. Mining activities involving blasting generate blast pressures on neighboring structures. The pressure distribution from a blasting source at a particular distance is considered nearly consistent over a normal reflecting surface. According to [26], a close-in explosion produces a pressure distribution that changes significantly in magnitude over the reflecting surface. This creates more complexity due to the non-uniform of pressure.

To determine the magnitude of peak overpressure, two major parameters are used: the charge weight and the distance between the blast source and the structure. By observing the pressure–time diagram depicted in Figure 4, two main phases can be identified. The positive part of the diagram is called the positive phase and has a duration t_o, as shown in Figure 4, while the negative part is called the negative phase and has a duration t_{o-}, also shown in Figure 4. According to [13], when a primary shock strikes a target, the reflected overpressure P_r instigates. The negative phase exists for a longer duration with lower intensity pressure than the positive phase. As the standoff distance increases, it can be noted that the duration/period of the positive blast wave phase increases, and that results in lower amplitude and a significantly longer-duration shock pulse.

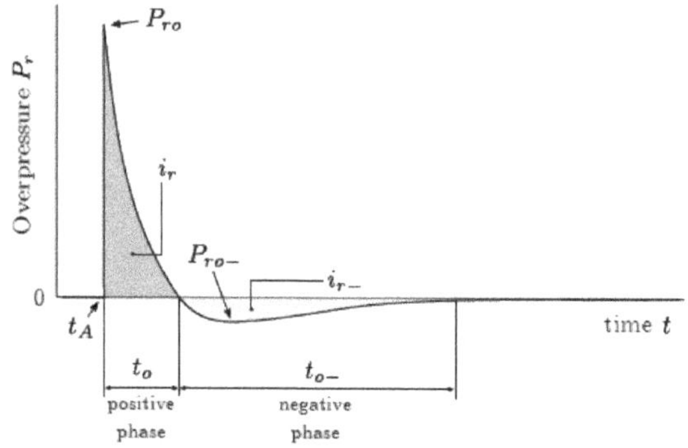

Figure 4. Shock wave distribution [27].

Using the Friedlander equation, the time evolution of the positive phase of the reflected pressure is analyzed (Friedlander, 1946):

$$P_r(t) = P_{ro}\left(1 - \frac{t^*}{t_o}\right)H[t^*](1 - H[t^* - t_o]\ H[t^*])\exp\left(-d\frac{t^*}{t_o}\right) \tag{5}$$

where *H[t*]* represents the step function, d is the exponential decay coefficient, and $t^* = t - t_A$, where t_A is depicted in Figure 4. According to Rigby et al. [28], the impulse i_{ro} or i_r associated with the positive phase, which symbolizes the area under the pressure curve, can be formulated as:

$$i_{ro} = \int_{t_A}^{t_A+t_0} P_r dt = \left[e^{-d} + d - 1\right]\frac{P_{ro}\ t_o}{d^2} \tag{6}$$

One of the most effective means of representing a blast impact is the use of the CONWEP model. According to [29], CONWEP is a model used to simulate the effects of a collection of conventional weapons, including air blast routines, breach, cratering, ground shock, and fragment and projectile penetration. The CONWEP charge property parameter is used in this study to simulate an air-based explosion using empirical data [30]. Furthermore, according to this consideration, a time history diagram of the pressure loading is built. In order to utilize this empirical model, one would need to define the equivalent TNT (trinitrotoluene) mass of the explosive as well as the source point (i.e., where the explosive is located). The initial process in calculating the explosive wave from a blast source other than TNT is to convert the charge mass to TNT equivalent mass [31].

Therefore, the CONWEP charge property is used in this study within commercial finite element software to simulate an air-based explosion by developing a time history pressure loading, similar to the one shown in Figure 4. The data, which were entered to define the blast charge properties, include the equivalent mass of TNT, a multiplication factor to convert from that mass unit into kilograms, and multiplication factors to convert from the standoff distance, time, or pressure to meter, second, or pressure in Pascals, respectively.

3.2. *Continuum Damage Law for the Masonry Units*

A concrete damaged plasticity law is used to represent damage on masonry units. Rate independence is claimed for this law, which is based on incremental plasticity theory. According to Lubliner et al. [32], Lee and Fenves [33], Tapkın et al. [34], and Daniel and Dubey [35], this constitutive description is appropriate for the analysis of quasi-brittle materials such as concrete and masonry. It relies on the concept of isotropic damaged elasticity for the representation of the irretrievable damage or failure that occurs during the cracking process for materials under fairly low pressure. The concrete damage plasticity law uses a non-associated potential plastic flow, which is in turn the implementation of the Drucker–Prager hyperbolic function for flow potential [36].

The common failure mechanisms that can be illustrated with this law are, namely, tensile cracking and compressive crushing. When unloading takes place, the elastic stiffness of the material is deemed damaged. This damage is implemented by introducing two damage variables as functions of the plastic strain, one for tension and the other for compression. A zero value of the damage variable indicates undamaged material, while a value equal to one indicates a total loss of strength. The corresponding uniaxial stress–strain relations, representing tension and compression, are provided below:

$$\sigma_t = (1 - d_t)E_0 (\varepsilon^t - \varepsilon^t_{pl}) \tag{7}$$

$$\sigma_c = (1 - d_c)E_0 (\varepsilon^c - \varepsilon^c_{pl}) \tag{8}$$

In the above equations, E_0 is the preliminary elastic stiffness of the material and d_t and d_c are the tensile and compressive damage variables, respectively.

The compressive and tensile stress–strain curves used in this work to define the compressive and tensile failure response of the masonry units on the numerical models, as well as the corresponding damage variables diagrams, are provided in the figures below. The uniaxial stress–strain behavior of concrete is modeled utilizing a Hognestad-type parabola [37], as per Figure 5 below. Figures 6–8 provide the compressive damage parameter as well as the tensile stress–strain law and the tensile damage parameter used for this model [37].

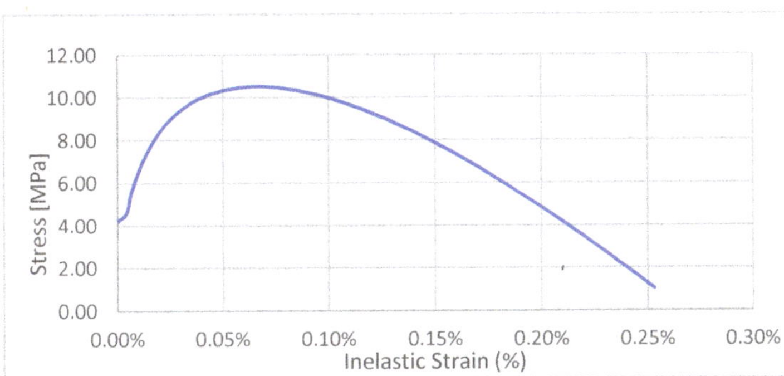

Figure 5. Compressive stress vs. strain diagram [37].

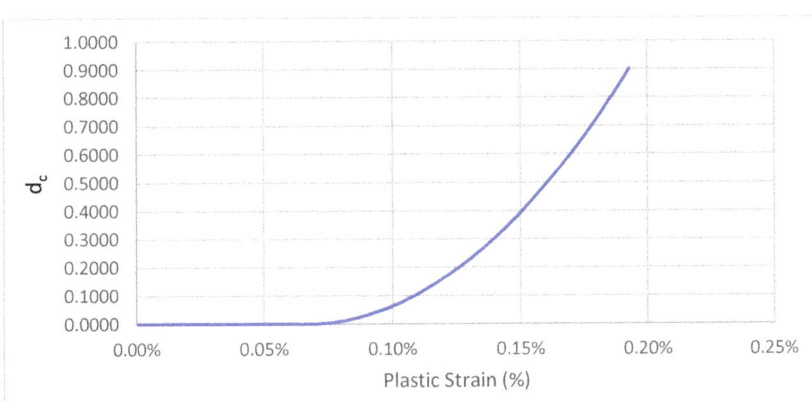

Figure 6. Compressive damage variable vs. plastic strain diagram [37].

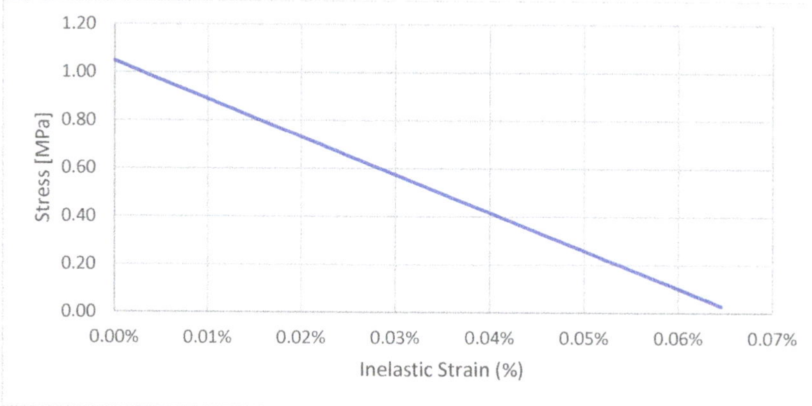

Figure 7. Tensile stress vs. strain diagram [37].

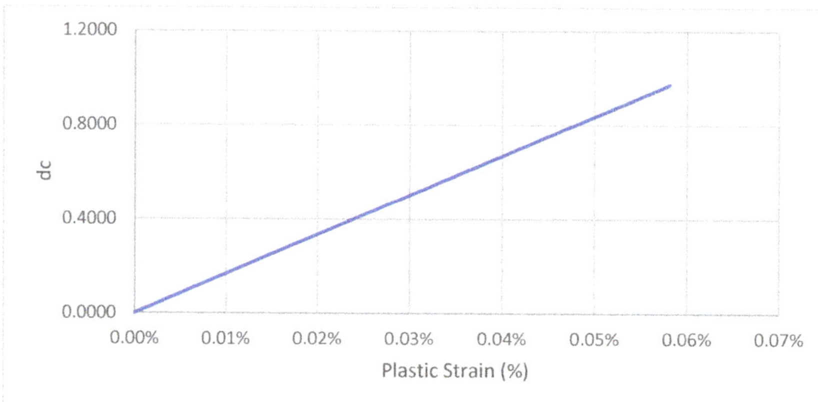

Figure 8. Tensile damage variable vs. plastic strain diagram [37].

The uniaxial tensile damage and uniaxial compressive damage parameters were developed using the post-failure stress as a function of cracking strain. The cracking strain is equal to the total strain minus the elastic strain of the undamaged material [32].

Some additional material properties used within the concrete damage plasticity law are provided in Table 1. The material properties for each masonry unit are provided in Table 2.

Table 1. Mechanical properties of the masonry unit and mortar [38].

Plasticity Parameter	Value
Dilation angle	30
Eccentricity parameter	0.1
Bi- and uni-directional compressive strength ratio	1.16
Stress ratio in tensile meridian	0.67
Viscosity parameter	0.001

Table 2. Material properties [38].

Material	Modulus of Elasticity (MPa)	Poisson's Ratio	Tensile Strength (MPa)	Compressive Strength (MPa)
Masonry Unit	15,500	0.15	1.05	10.5

3.3. The Geometry of the Masonry Walls

The dimensions of each masonry unit considered in this study are equal to 430 mm × 140 mm × 190 mm. The size of each unit is as per the Concrete Manufactures Association [39]. Low-cost housing in South Africa often uses concrete masonry blocks and clay bricks. This paper focuses on the use of concrete blocks, and the following limitations are noted:

- A single-leaf wall is considered, and the wall is unreinforced.
- Category 1 buildings [40].

Two geometries are used in this study for the walls, as shown in Figures 9 and 10. The first is a solid wall and the second represents a wall with an opening.

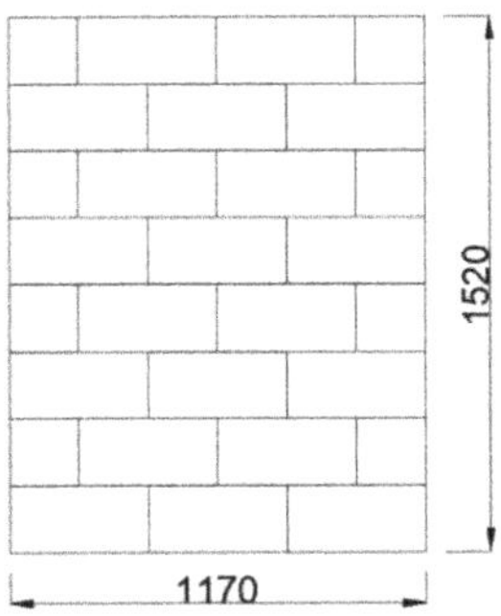

Figure 9. Geometry 1 (wall with no opening (mm)).

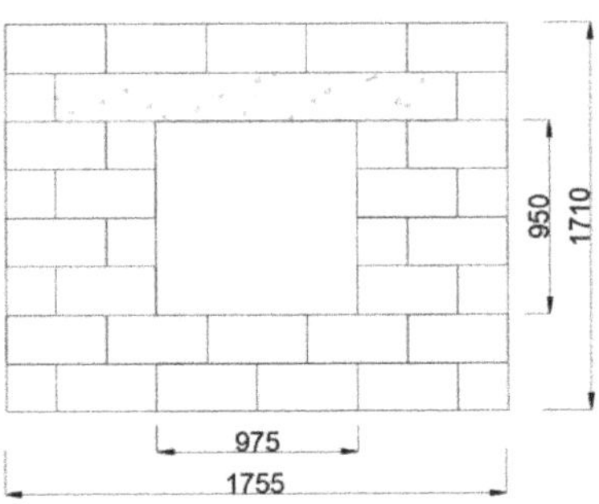

Figure 10. Geometry 2 (wall with an opening (mm)).

3.4. Details of the Finite Element Model

Figure 11 shows the mesh that is adopted in this study for the models without and with an opening. Three-dimensional, eight-node linear brick elements are used, with the element side equal to 40 mm for both walls. A total number of 4800 elements for the model without the opening and 5600 elements for the model with the opening are used, as shown in Figure 11.

(**a**) (**b**)

Figure 11. Mesh of the considered masonry walls: (**a**) with no opening and (**b**) with a window opening.

All four sides on the perimeter of each of the two walls are considered as fixed in three translational degrees of freedom, according to the coordinate system shown in Figure 11. It is noted that the restraining of the top side of the walls in the Z-direction is attributed to the assumption that an upper slab or roof will provide restraint in that direction.

Concerning the loading of the models, two load steps are used. In an initial, pre-existing step, a vertical pressure of 0.25 MPa is applied to the top side of the structure. In the first load step, a horizontal shear (in-plane) displacement of 10 mm is applied to the top side of the walls. Alternatively, the wall with no horizontal in-plane displacement is also considered. In the second load step, the blast loading is applied.

The simulation is conducted using explicit dynamic analysis. This type of analysis is appropriate since it is able to capture the very short duration of the blast action. It is noted that the explicit dynamic analysis was originally developed to simulate high-speed dynamic events that would otherwise require significant computational resources within implicit codes. For the implementation of this analysis, an automatic time incrementation is used.

For the application of the contact–friction conditions between the masonry blocks, the method of Lagrange multipliers is used. A friction coefficient equal to 0.45 is assigned to the interfaces.

It is noted that for the implementation of the blast load, a charge weight expressed in TNT at the standoff distances of 100 m, 50 m, and 20 m is used. In addition, the effect of the blast weight, as well as the effect of changing the blast charge while keeping the distance constant and changing the standoff distance while keeping the blast charge weight constant, are also investigated. Only the front surface of the walls is loaded (incident surface). In the following sections, results obtained from various parametric investigations, emphasizing the corresponding failure mechanisms, are provided.

4. Validation of the Proposed Model

Before presenting the results from the simulations, a validation of the proposed model is conducted by comparing the output with published numerical studies. Then, the results of the investigation are elaborated, emphasizing collapse mechanisms for the used load cases.

A comparison between results derived from the proposed model and existing numerical solutions is provided. As depicted in the figures below, similar collapse mechanisms were obtained when a blast load or in-plane, vertical pressure, and shear displacement loads are applied to the wall.

- Masi et al. [13] depicted complete failure, indicating that after a wall is damaged close to the boundary, it comes out in one piece in the middle portion. The explosive weight was 810 kg at 37 m standoff distances (Figure (a)). Out-of-plane experiments conducted by Du et al. [41] depicted complete failure in the middle of a real wall (Figure (b)).

(a) (b) (c)

The proposed model depicted complete failure of the middle section of the wall, under a blast load with an explosive weight of 810 kg at a standoff distance of 37 m. This is shown in Figure (c).

- Salmanpour [21] predicted sliding failure along staircase-shaped cracks as mode of failure for a wall subjected to vertical pressure and shear displacement loading. The damage to buildings during the 2002 Molise earthquake in Italy, as reported in [42], resembles the crack pattern shown below in Figure (d).

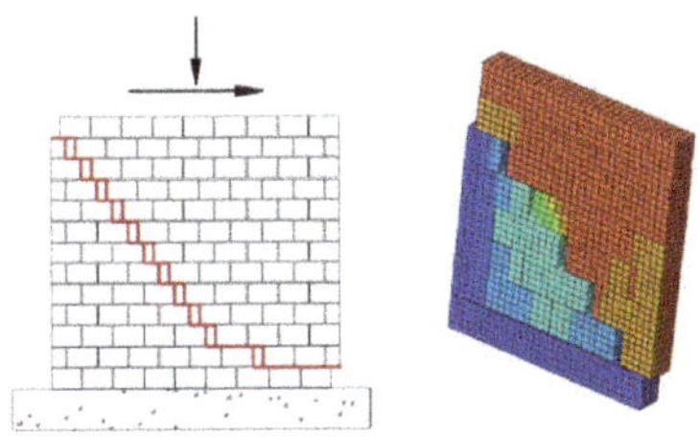

(d) (e)

The proposed model depicted staircase-shaped cracks when subjected to vertical pressure and shear displacement loading. This is shown in Figure (e).

5. Results Obtained from the Proposed Numerical Scheme

In the next sections, results depicting the structural response of two masonry walls, one without an opening and one with an opening, are provided. Within this investigation, parametric studies depicting the influence of the variation in the weight of charge and the blast source–structure distance on the response of the walls, were conducted. Relevant discussions emphasize the collapse modes, which arise in the provided framework. In total, the cases listed in Table 3 are considered for the walls without and with an opening.

Table 3. Cases considered in the framework of this investigation.

Case Description	Blast Load	Standoff Distance
Case 1 (solid wall loaded with vertical pressure, horizontal shear displacement, and blast load).	100 kg TNT	20 m, 50 m
	200 kg TNT	20 m, 50 m
	1150 kg TNT	20 m, 50 m, 100 m
Case 2 (wall with an opening loaded with vertical pressure, horizontal shear displacement, and blast load).	1150 kg TNT	50 m
	2000 kg TNT	5 m
	3500 kg TNT	10 m
Case 3 (solid wall loaded with vertical pressure and blast load). No horizontal shear displacement is considered.	100 kg TNT	50 m
	200 kg TNT	100 m
	1150 kg TNT	100 m

5.1. Case 1: Solid Masonry Wall Loaded with Vertical Pressure, Shear Displacement, and Blast Load

This section provides results obtained from simulations on a solid wall loaded at its top boundary surface with a vertical downward pressure and a horizontal shear displacement load. A blasting action is also applied to the wall.

Figure 12 depicts the displacement and the failure mechanism, which are obtained from an explosive weight of 100 kg at a standoff distance of 20 m. According to this figure, in-plane diagonal cracking in the form of opening/sliding between the blocks is accompanied by some out-of-plane flexural displacement, attributed to the blast load. It can be observed that though the wall does not collapse totally, significant displacements are developed. In addition, according to Figure 12b,c, significant compressive and tensile failure is developed on the masonry blocks. This is attributed to the out-of-plane flexural displacement of the fixed (in its perimeter) wall. As expected, tensile failure is more expanded in the wall than compressive failure.

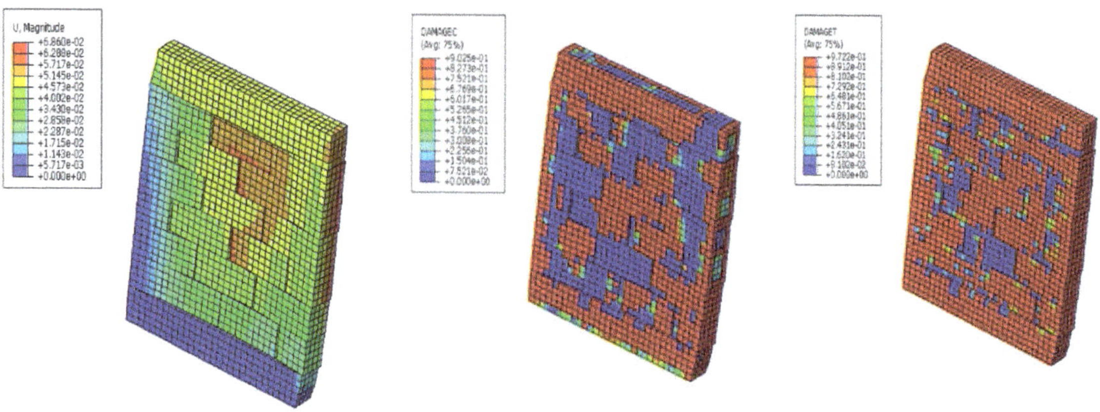

(a) (b) (c)

Figure 12. Solid wall loaded with a charge weight of 100 kg at a distance of 20 m: (**a**) displacement of the wall at the end of the simulation (m), (**b**) compressive damage variable, and (**c**) tensile damage variable distribution.

Next, the weight of the explosive is gradually increased to 200 kg (Figure 13) and 1150 kg (Figure 14) while the standoff distance is kept constant and equal to 20 m. As observed in Figure 13a, the out-of-plane deflection becomes higher compared with Figure 12a, due to the increase in the blast weight. In addition, Figure 13b,c shows that extensive compressive and tensile failure is developed in the masonry blocks. Figure 14 shows that for the maximum quantity of explosive weight, out-of-plane deflection dominates the in-plane cracking. In addition, the wall fails completely for this maximum explosive weight when it is considered at the same distance of 20 m, as above.

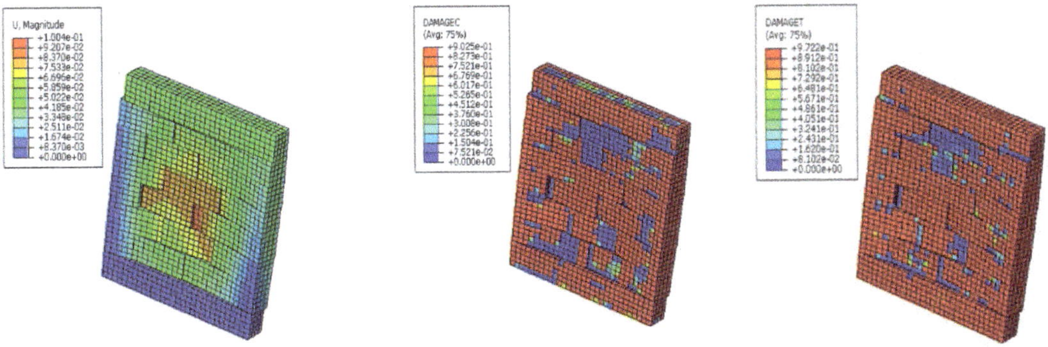

(a) (b) (c)

Figure 13. Solid wall loaded with a charge weight of 200 kg at a distance of 20 m: (**a**) displacement of the wall at the end of the simulation (m), (**b**) compressive damage variable, and (**c**) tensile damage variable distribution.

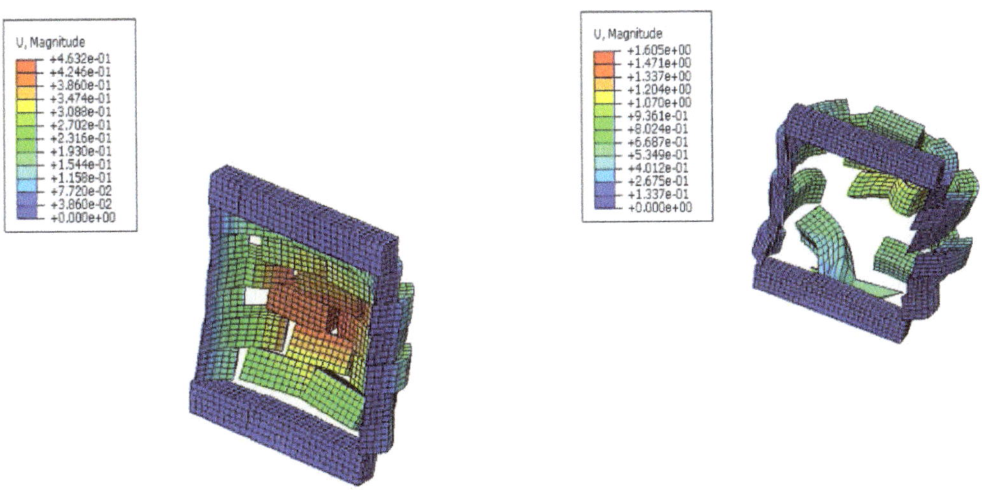

(a) (b)

Figure 14. Displacement of the solid wall loaded with a charge weight of 1150 kg at a distance of 20 m: (**a**) depicted in a previous time step prior to complete damage of the wall and (**b**) depicted at the final time step.

Next, an investigation using an increased standoff distance of 50 m is conducted, using explosive weights of 100 kg, 200 kg, and 1150 kg, respectively. From the first case of 100 kg explosive weight, it is determined that the in-plane failure mode is dominant, as shown in Figure 15. A diagonal cracking appears, in this case, at the top part of the wall. On the contrary, for the lower distance of 20 m shown in Figure 12, both in-plane and out-of-plane flexural deflection are observed.

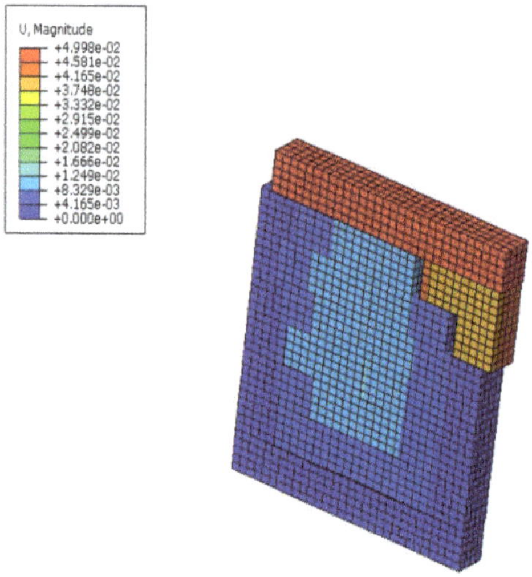

Figure 15. Displacement of the solid wall loaded with a charge weight of 100 kg at a distance of 50 m.

For the case of 200 kg explosive weight, the failure mechanism which arises is a stair-case diagonal cracking along the wall, as shown in Figure 16. This indicates that the failure mode changes compared with Figure 13, where mainly an out-of-plane response is observed for the same explosive weight and lower standoff distance (20 m). It is noted that the damage pattern depicted in Figure 16 is observed in masonry walls that are loaded with in-plane actions. Some limited out-of-plane flexural displacement also arises in this case.

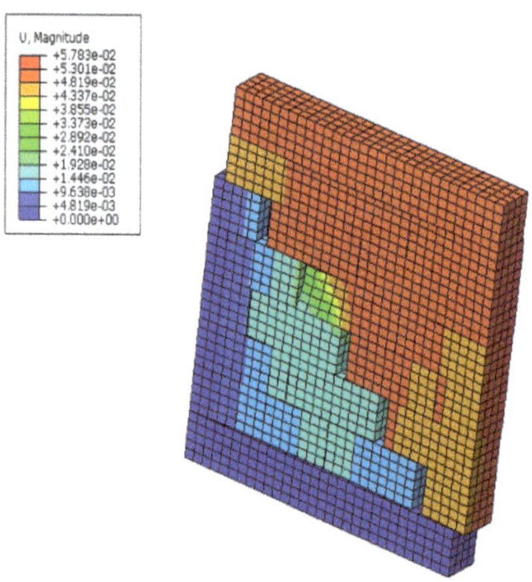

Figure 16. Displacement of the solid wall loaded with a charge weight of 200 kg at a distance of 50 m.

For the case of the highest explosive weight of 1150 kg (standoff distance 50 m), it is determined that out-of-plane flexural deflection is the dominant response, leading to a corresponding failure mode. Moreover, as can be observed in Figure 17b,c, both tensile and compressive failure are developed in the whole mass of the wall. By comparing this with the case of the same explosive weight and lower standoff distance (20 m) shown in Figure 14, it is noticed that although both walls fail due to the out-of-plane response attributed to the blast action, the model with the lower standoff distance leads to a total out-of-plane collapse.

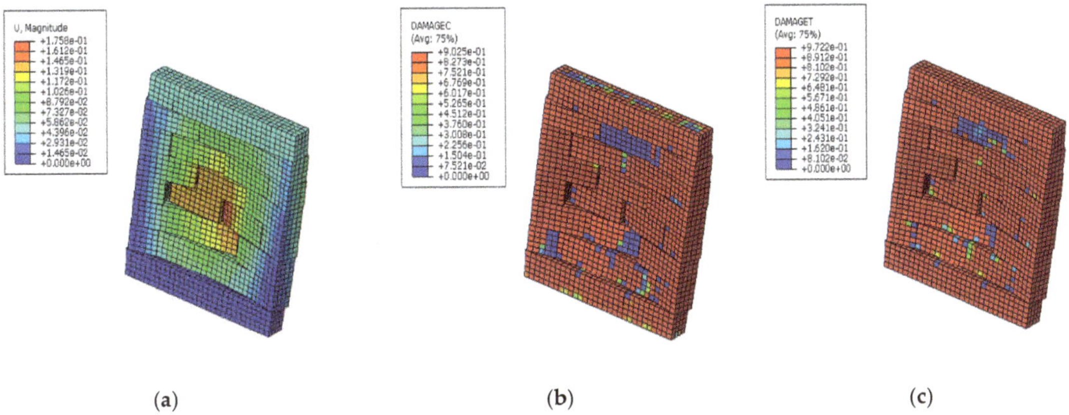

Figure 17. Solid wall loaded with a charge weight of 1150 kg at a distance of 50 m: (**a**) displacement of the wall at the end of the simulation (m), (**b**) compressive damage variable, and (**c**) tensile damage variable distribution.

When an increase in the standoff distance from 50 m to 100 m is considered for the case of the maximum explosive weight of 1150 kg, the failure mode changes and diagonal in-plane cracking becomes dominant, as shown in Figure 18, contrary to the out-of-plane flexural deflection observed at a distance of 50 m (Figure 17). Some out-of-plane flexural deflection, accompanied by tensile failure at the perimeter and at the central part of the wall, is also obtained, as shown in Figure 18. Compressive failure is more limited and is mainly developed at the wall's bottom corner.

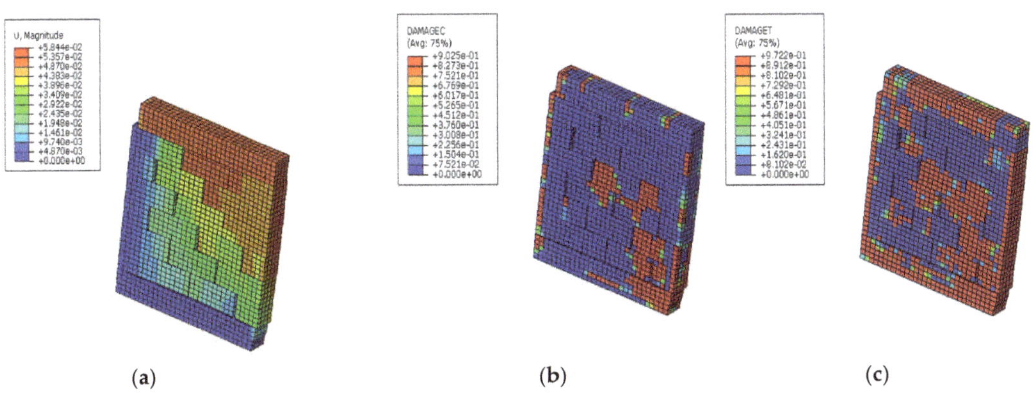

Figure 18. Solid wall loaded with a charge weight of 1150 kg at a distance of 100 m: (**a**) displacement of the wall at the end of the simulation (m), (**b**) compressive damage variable, and (**c**) tensile damage variable distribution.

To summarize the effect of varying standoff distances for each explosive weight, the diagrams shown in Figure 19 are used. It is observed that for bigger explosive weights and lower standoff distances, higher deflections of the wall are obtained. When the standoff distance is increased, the impact of the blast loading on the structural system is reduced since the maximum displacements are also reduced. It is noted that for the case of out-of-plane response for the maximum explosive weight and minimum distance, a large deflection is obtained, as depicted with point X in the graph shown in Figure 19a. This value is only indicative, highlighting a total collapse of the central part of the wall.

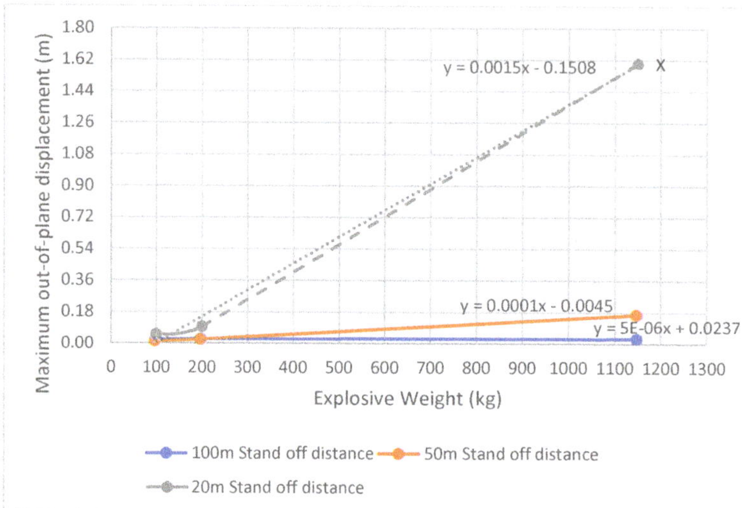

(a)

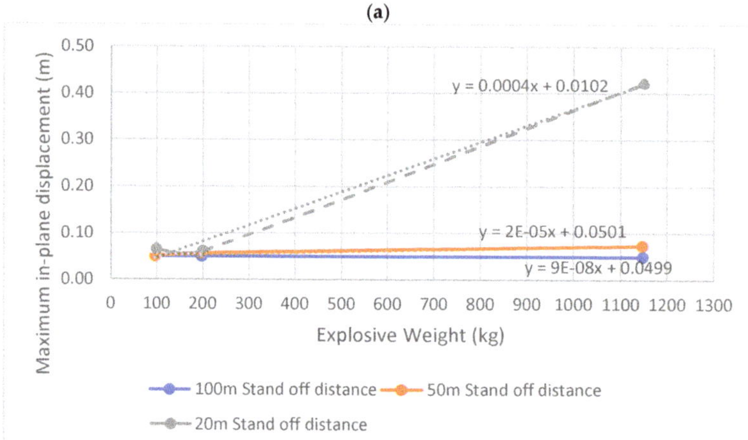

(b)

Figure 19. Displacement–explosive weight diagrams depicting the (**a**) out-of-plane and (**b**) in-plane response.

In Figure 19a,b, linear regression formulas are determined, providing approximate mathematical expressions that can be used to calculate the deflection of the walls for different explosive weight values and standoff distances. In particular, for standoff distances of 20 m, 50 m, and 100 m, the correlation coefficients are found to be equal to $R^2 = 0.9962$, $R^2 = 0.9993$, and $R^2 = 0.85$, respectively, for the out-of-plane response. In terms of the in-plane response, standoff distances 20 m, 50 m, and 100 m provide correlation coefficients of 0.9906, 0.9618, and 0.9926, respectively. It is noted that an R^2 close to 1 indicates that the regression prediction is of satisfactory accuracy. Equations on the graphs were tested by inserting a random independent variable "x" (explosive weight) to estimate the displacement "y" (deflection of the wall).

5.2. Case 2: Masonry Wall with an Opening Loaded with Vertical Pressure, Shear Displacement, and Blast Load

To capture the influence of openings (windows) on the structural response of masonry walls under blast actions, a new model is developed, introducing a window in the middle of the wall, as shown in Figure 10. Similar to Section 5.1, a vertical downward pressure and a shear in-plane loading applied at the top surface of the wall are considered together with the blast action. To simulate the influence of the lintel, which is a concrete beam usually built just above the opening and in contact with the masonry above, the vertical displacements for the elements above the window are restricted.

In Figure 20, the displacement contour plot, which was obtained when an explosive weight of 1150 kg at a standoff distance of 50 m is applied to the wall, is shown. It is observed that relatively low max displacements arise at the final load step, with an in-plane diagonal cracking above the window on the left-hand side of the wall just emerging. By comparing this with Figure 17, which depicts the response of the solid wall under the same blast loading and standoff distance, it appears that the wall with the opening develops significantly lower deflection with no obvious out-of-plane deformation, contrary to the solid wall. This is attributed to the fact that the blast load is modeled as a surface force under the incident wave (CONWEP), and the damage or effect of the blast is directly proportional to the exposed surface. In addition, the window is inserted at the middle part of the wall, where the out-of-plane response due to the blast load would become maximum in the case of a solid wall.

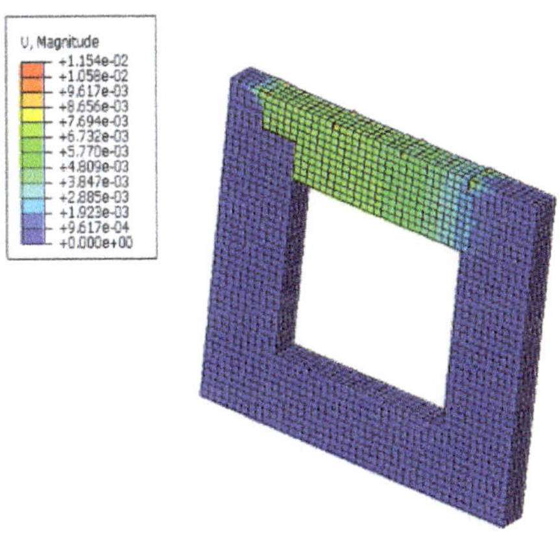

Figure 20. Displacement contour plot at the end of the simulation for a wall with an opening subjected to 1150 kg TNT at a standoff distance of 50 m.

Next, to depict the failure of the wall with the opening, the weight of the explosive is increased and the standoff distance is reduced. In particular, the displacement plots for the wall obtained from explosive weights of 2000 kg and 3500 kg at standoff distances of 5 m and 10 m are provided in Figure 21. For both cases, out-of-plane failure is obtained as depicted in Figure 21a,b with the case of 2000 kg explosive weight at a distance of 5 m being the most severe.

In an effort to provide a qualitative comparison of the response of the walls with and without a window, it is observed that the most severe collapse of the wall with the window occurs for 2000 kg explosive weight positioned at 5 m (Figure 21a). The corresponding most severe collapse of the solid wall occurs for 1150 kg explosive weight positioned at 20 m (Figure 14). Thus, it seems that the opening in the middle of the wall significantly reduces the effects of the blast action. This observation is in agreement with the findings provided in [43], where the area of an opening in a masonry building is shown to have a significant impact on both the normal and shear stresses produced by blast.

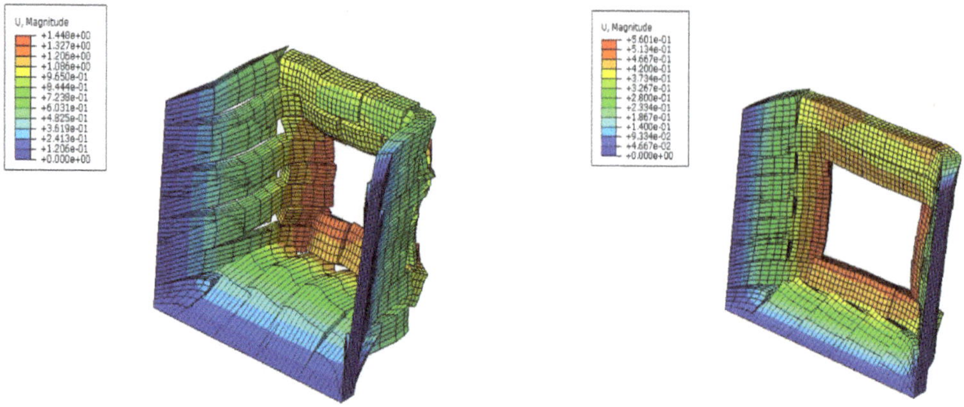

(**a**) (**b**)

Figure 21. Displacement contour plot at the end of the simulation for a wall with an opening subjected to (**a**) 2000 kg TNT at a standoff distance of 5 m and (**b**) 3500 kg TNT at a standoff distance of 10 m.

5.3. Case 3: Solid Masonry Wall Loaded with Vertical Pressure and Blast Load

In this section, the response of the wall without an opening loaded with a vertical downward pressure at the top surface and the blast action is investigated. Thus, contrary to Section 5.1 (and Section 5.2), no shear horizontal displacement is applied at the top of the wall. In Figure 22, the response of the wall subjected to 100 kg and 200 kg TNT at standoff distances of 50 m and 100 m, respectively, is provided. In Figure 23, the response of the wall subjected to 1150 kg TNT and 100 m standoff distance is shown.

In both Figures 22 and 23, the out-of-plane deflection characterizes the response of the wall. Contrary to this behavior, the model of the wall loaded with shear horizontal displacement, 100 kg explosive weight, and 50 m standoff distance, resulted in diagonal in-plane cracking (Figure 15) with significantly higher maximum deflection, compared with the wall shown in Figure 22a.

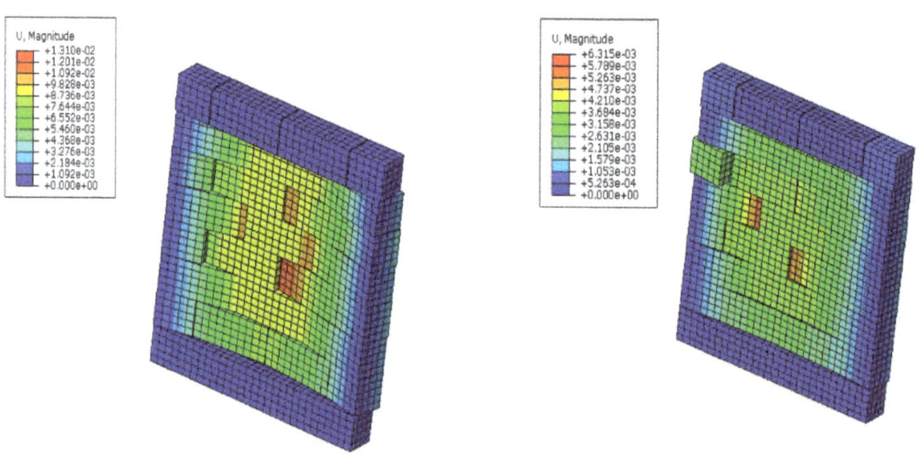

(**a**) (**b**)

Figure 22. Solid wall subjected to (**a**) 100 kg TNT at a standoff distance of 50 m and (**b**) 200 kg TNT at a standoff distance of 100 m when no horizontal displacement loading is applied to the top surface of the wall (a scale factor equal to 10 is used to magnify the displacement contour plots).

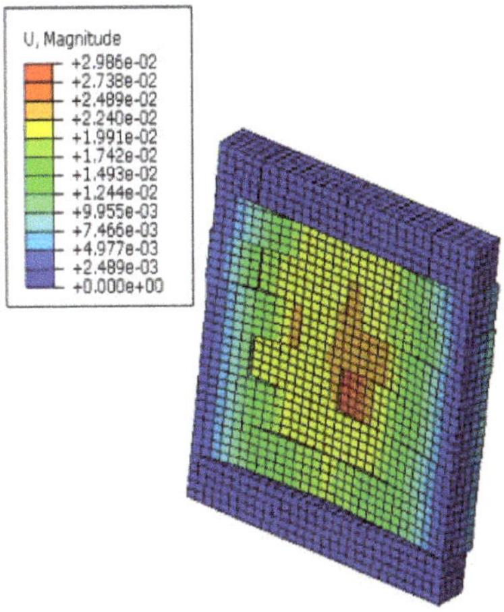

Figure 23. Solid wall subjected to 1150 kg TNT at a standoff distance of 100 m when no horizontal displacement loading is applied to the top surface of the wall (a scale factor equal to 10 is used to magnify the displacement contour plots).

A similar comparison can be made between Figures 18 and 23, representing the displacement contour plot of the wall under the same blast load and standoff distance (1150 kg, 100 m) with and without the shear displacement load, respectively. As shown in Figure 18, a significantly higher maximum deflection is obtained when the shear displacement load is applied to the wall, compared with Figure 23. In addition, the in-plane diagonal cracking is dominant in Figure 18.

6. Conclusions

In this article, the response of masonry walls under static in-plane and blast loads is investigated using non-linear finite element analysis software [44]. For the simulation of damage in the interfaces between the stone blocks, unilateral contact–friction interfaces are applied to depict opening and sliding failure. In addition, a concrete damage plasticity model is used to describe tensile and compressive damage in the blocks. The proposed scheme is applied to a solid masonry wall and to a wall with an opening (window).

This investigation aims in highlighting potential collapse mechanisms by testing different blast load parameters, namely, the weight of the explosive and the standoff distance between the source of the explosion and the structure. The influence of a horizontal shear displacement in-plane loading at the top of the wall is also investigated.

According to the findings of this study, the failure mode of the wall loaded with both shear in-plane displacement and the blast action can be either in-plane diagonal cracking or out-of-plane flexural failure. The first mode arises when the shear in-plane displacement is the dominant loading, compared with the blasting action, while the second arises when the blast is the dominant loading. For the same material properties and wall dimensions, the weight of the explosive and the standoff distance are the critical parameters, which determine which of the two loading types dominates. In the results section, case studies highlighting both failure modes are discussed for various values of the explosive weight and the standoff distance. A combination of both failure modes can also arise, depending on the values of these parameters.

Another outcome of this work is the fact that the presence of an opening (window) in the wall may reduce the effect of the blast action by decreasing the out-of-plane response of the structure. The reason for this is that due to the opening being located in the middle of the wall, the blast load is not applied to this critical (for out-of-plane flexure) middle part of the surface of the wall. Thus, this study shows that the blast action must occur at a closer standoff distance compared with the solid wall, in order to cause significant damage to the structure.

When no shear displacement in-plane loading is applied, the response is dominated by the out-of-plane flexural deflection, attributed to the blasting action. In this case, lower maximum displacements are obtained compared with the wall loaded with shear displacement and blast actions.

Several future investigations could be used to extend the present work. A potential concept is to study the influence of the area, position, and number of windows on the response of the walls under blast actions. The usage of different initial static loading could modify the results, as was shown in the conducted numerical investigation. Design or re-design based on these findings could also form an interesting research topic. Another concept is related to the implementation of data-driven structural dynamics, introducing machine learning tools, to evaluate the influence of several parameters such as the dimensions of the walls and the blast load parameters on their structural response.

Author Contributions: Conceptualization, G.A.D.; methodology, G.A.D.; formal analysis, S.G.T.; investigation, S.G.T. and G.A.D.; resources, G.A.D.; data curation, S.G.T.; writing—original draft preparation, S.G.T.; writing—review and editing, S.G.T., G.A.D. and G.E.S.; visualization, S.G.T.; supervision, G.A.D. and G.E.S.; project administration, G.A.D. All authors have read and agreed to the published version of the manuscript.

Funding: This research received no external funding.

Data Availability Statement: Not applicable.

Conflicts of Interest: The authors declare no conflict of interest.

References

1. Ahmad, S.; Elahi, A.; Pervaiz, H.; Rahman, A.G.A.; Barbhuiya, S. Experimental study of masonry wall exposed to blast loading. *Mater. Constr.* **2014**, *64*, 2–6. [CrossRef]
2. Gertsch, L.; Baird, J. *A Phased Array Approach to Rock Blasting*; University of North Texas, UNT Digital Library: Denton, TX, USA, 2006; pp. 13–40. Available online: https://digital.library.unt.edu (accessed on 22 March 2022).
3. Aloui, M.; Bleuzen, Y.; Essefi, E.; Abbes, C. Ground Vibrations and Air Blast Effects Induced by Blasting in Open Pit Mines: Case of Metlaoui Mining Basin, Southwestern Tunisia. *J. Geol. Geophys.* **2016**, *5*, 5–7. [CrossRef]
4. Ipmawati, M.R.; Nainggolan, D.R.; Wiyono, B.; Sunary, R. Effect double-primer placement for improving the fragmentation on harder material in stemming column: A case study. *IOP Conf. Ser. Earth Environ. Sci.* **2018**, *212*, 012064. [CrossRef]
5. Hao, D. Numerical Modelling of Masonry Wall Response to Blast Loads. *Aust. J. Struct. Eng.* **2009**, *10*, 37–52. [CrossRef]
6. Wang, W.; Zhang, D.; Fangyun, L.; Wang, S.C.; Tang, T. Experimental study and numerical simulation of the damage mode of a square reinforced concrete slab under close-in explosion. *Eng. Fail. Anal.* **2013**, *27*, 41–51. [CrossRef]
7. Zhao, C.F.; Chen, J.Y. Damage mechanism and mode of square reinforced concrete slab subjected to blast loading. *Theor. Appl. Fract. Mech.* **2013**, *63–64*, 54–62. [CrossRef]
8. Macorini, L.; Izzuddin, B.A. Nonlinear Analysis of Unreinforced Masonry Walls under Blast Loading Using Mesoscale Partitioned Modeling. *J. Struct. Eng.* **2014**, *140*, A4014002. [CrossRef]
9. Kernicky, T.P.; Whelan, M.J.; Weggel, D.C.; Rice, C.D. Structural Identification and Damage Characterization of a Masonry Infill Wall in a Full-Scale Building Subjected to Internal Blast Load. *J. Struct. Eng.* **2015**, *141*, D4014013. [CrossRef]
10. Jia, H.; Yu, L.; Wu, G. Damage Assessment of Two-Way Bending RC Slabs Subjected to Blast Loadings. *Publ. Corp. Sci. World J.* **2014**, *2014*, 718702. [CrossRef]
11. D'Altri, A.M.; Miranda, S.; Castellazzi, G.; Sarhosis, V. A 3D Detailed Micro-Modelling Approach for the In-Plane and Out-Of-Plane Analysis of Masonry Structures. *Comput. Struct.* **2018**, *206*, 18–30. [CrossRef]
12. Lourenço, P.B.; Rots, J.G.; Blaauwendraad, J. Two Approaches for the Analysis of Masonry Structures: Micro and Macro-Modeling. *Heron* **1995**, *40*, 313–338.
13. Masi, F.; Stafanou, I.; Maffi-Berthier, V.; Vannucci, P. A Discrete Element Method based-approach for arched masonry structures under blast loads. *Eng. Struct.* **2020**, *216*, 110721. [CrossRef]
14. Shamim, S.; Shamd, S.; Khan, A. Finite element analysis of masonry wall subjected to blast loading. *Int. J. Adv. Mech. Civ. Eng.* **2019**, *6*, 50–53.
15. Shamim, S.; Shamd, S.; Khan, A. Numerical Modelling of Masonry Panel Subjected to Surface Blast Loading. *J. Xi'an Univ. Archit. Technol.* **2020**, *12*, 846–856.
16. Abdulla, K.; Cunningham, L.; Gillie, M. Non-linear simulation of masonry behaviour under cyclic loads. In Proceedings of the 2017 MACE PGR Conference, Manchester, UK, 3 April 2017; pp. 20–22.

17. Anas, S.M.; Alam, M.; Umair, M. Strengthening of braced unreinforced brick masonry wall with (i) C-FRP wrapping, and (ii) steel angle-strip system under blast loading. *Mater. Today Proc.* **2022**, *58*, 1181–1198. [CrossRef]
18. Shariq, M.; Alam, M.; Husain, A.; Islam, N. Response of strengthened unreinforced brick masonry wall with (1) mild steel wire mesh and (2) CFRP wrapping, under close-in blast. *Mater. Today Proc.* **2022**, *64*, 643–654. [CrossRef]
19. Shamim, S.; Khan, R.A.; Ahmad, S. Fragility analysis of masonry wall subjected to blast loading. *Structures* **2022**, *39*, 1016–1030. [CrossRef]
20. Yu, Q.; Zeng, D.; Xu, X.; Li, S.; Dong, W.; Dai, L. Experimental and numerical investigation of polymer-reinforced and normal autoclaved aerated concrete masonry walls under large TNT explosive loads. *Int. J. Impact Eng.* **2022**, *164*, 104188. [CrossRef]
21. Salmanpour, A.H. Displacement Capacity of Structural Masonry. Ph.D. Thesis, ETH Zurich, Zurich, Switzerland, 2017.
22. De Villiers, W.I. Computational and Experimental Modelling of Masonry Walling towards Performance-Based Standardisation of Alternative Masonry Units for Low-Income Housing. Ph.D. Thesis, Stellenbosch University, Stellenbosch, South Africa, 2019.
23. Kömürcü, S.; Gedikli, A. Macro and Micro Modeling of the Unreinforced Masonry Shear Walls. *Eur. J. Eng. Nat. Sci.* **2019**, *3*, 116–123.
24. Ranji, A.R.; Esmaeli, A. Blast load response of one-way reinforced concrete slabs retrofitted with Fiber reinforced plastic. *Proc. Odessa Polytech. Univ.* **2018**, *2*, 49–58. [CrossRef]
25. Tan, W.A.; Tingatinga, E.A.; Alvarez, V. Blast load analysis and simulation of unreinforced concrete masonry. *J. Phys. Conf. Ser.* **2019**, *1264*, 012008. [CrossRef]
26. Weggel, D.C. Blast threats and blast loading. In *Blast Protection of Civil Infrastructures and Vehicles Using Composites*; Series in Civil and Structural Engineering; Woodhead Publishing: Swaston, UK, 2010; pp. 3–43. [CrossRef]
27. Botez, M.D.; Bredean, L.A. Numerical Study of a RC Slab Subjected to Blast: A Coupled Eulerian-Lagrangian Approach. *IOP Conf. Ser. Mater. Sci. Eng.* **2019**, *471*, 2–5. [CrossRef]
28. Rigby, S.E.; Tyas, A.; Bennett, T.; Clarke, S.D.; Fay, S.D. The Negative Phase of the Blast Load. *Int. J. Prot. Struct.* **2014**, *5*, 1–19. [CrossRef]
29. Alsayed, S.H.; Elsanadedy, H.M.; Al-Zaheri, Z.M.; Salloum, Y.A.; Abbas, H. Blast response of GFRP strengthened, infill masonry walls. *Constr. Build. Mater.* **2016**, *115*, 438–451. [CrossRef]
30. Bolhassani, M.; Hamid, A.A.; Lau, A.C.W.; Moon, F. Simplified micro modelling of partially grouted masonry assemblages. *Constr. Build. Mater.* **2015**, *83*, 159–173. [CrossRef]
31. Draganić, H.; Sigmund, V. Blast Loading On Structures. *Tech. Gaz.* **2012**, *19*, 643–652.
32. Lubliner, J.; Oliver, J.; Oller, S.; Oñate, E. A plastic-damage model for concrete. *Int. J. Solids Struct.* **1989**, *25*, 299–329. [CrossRef]
33. Lee, J.; Fenves, G.L. Plastic-Damage Model for Cyclic Loading of Concrete Structures. *J. Eng. Mech.* **1998**, *124*, 892–900. [CrossRef]
34. Tapkin, S.; Tercan, E.; Motsa, S.; Drosopoulos, G.; Stavroulaki, M.; Maravelakis, E.; Stavroulakis, G. Structural Investigation of Masonry Arch Bridges Using Various Nonlinear Finite-Element Models. *J. Bridge Eng.* **2022**, *27*, 04022053. [CrossRef]
35. Daniel, J.; Dubey, R. Finite Element Simulation of Earthquake Resistant Brick Masonry Building Under Shock Loading. *Adv. Struct. Eng.* **2014**, *81*, 1027–1038. [CrossRef]
36. Dogu, M.; Menkulasi, F. A flexural design methodology for UHPC beams posttensioned with unbonded tendons. *Eng. Struct.* **2020**, *207*, 110193. [CrossRef]
37. Hognestad, E. *A Study of Combined Bending and Axial Load in Reinforced Concrete Members*; Bulletin Series No.399; University of Illinois, Engineering Experiment Station: Chicago, IL, USA, 1951; pp. 32–54.
38. Iuorio, O.; Dauda, J.A. Retrofitting Masonry Walls against Out-Of-Plane Loading with Timber Based Panels. *Appl. Sci.* **2021**, *11*, 5443. [CrossRef]
39. Concrete Manufactures Association. *Concrete Masonry Manual*, 9th ed.; Concrete Manufactures Association: Menlo Park, CA, USA, 2011. Available online: https://www.cma.org.za/Portals/0/Docs/brickasandblocks (accessed on 22 April 2022).
40. *SANS 10400-A 2010*; The Application of the National Building Regulations Part A: General Principles and Requirements, 3rd ed. SABS Standards Division: Pretoria, South Africa, 2010; ISBN 978-0-626-25157-4.
41. Wei, J.; Du, Z.; Zheng, Y.; Ounhueane, O. Research on Damage Characteristics of Brick Masonry under Explosion Load. *Shock Vib.* **2021**, *2021*, 5519231. [CrossRef]
42. Decanini, L.; De Sortis, A.; Goretti, A.; Langenbach, R.; Mollaioli, F.; Rasulo, A. Performance of masonry buildings during the 2002 Molise, Italy, earthquake. *Earthq. Spectra* **2004**, *20* (Suppl. S1), 191–220. [CrossRef]
43. Raza, I.; Shamim, S.; Ahmad, S.; Khan, R.A. Analysis of two-storey masonry structure under blast loading. *Mater. Today Proc.* **2023**, *80*, 1605–1610. [CrossRef]
44. ABAQUS v. 6.14.2 Users Manual. Available online: http://130.149.89.49:2080/v2016/index.html (accessed on 13 March 2022).

computation

MDPI

Article

The MLDAR Model: Machine Learning-Based Denoising of Structural Response Signals Generated by Ambient Vibration

Spyros Damikoukas and Nikos D. Lagaros *

Institute of Structural Analysis and Antiseismic Research, School of Civil Engineering, National Technical University of Athens, 9 Heroon Polytechniou Str., Zografou Campus, GR-15780 Athens, Greece; spydamik@gmail.com

* Correspondence: nlagaros@central.ntua.gr; Tel.: +30-210-772-2625

Abstract: Engineers have consistently prioritized the maintenance of structural serviceability and safety. Recent strides in design codes, computational tools, and Structural Health Monitoring (SHM) have sought to address these concerns. On the other hand, the burgeoning application of machine learning (ML) techniques across diverse domains has been noteworthy. This research proposes the combination of ML techniques with SHM to bridge the gap between high-cost and affordable measurement devices. A significant challenge associated with low-cost instruments lies in the heightened noise introduced into recorded data, particularly obscuring structural responses in ambient vibration (AV) measurements. Consequently, the obscured signal within the noise poses challenges for engineers in identifying the eigenfrequencies of structures. This article concentrates on eliminating additive noise, particularly electronic noise stemming from sensor circuitry and components, in AV measurements. The proposed MLDAR (Machine Learning-based Denoising of Ambient Response) model employs a neural network architecture, featuring a denoising autoencoder with convolutional and upsampling layers. The MLDAR model undergoes training using AV response signals from various Single-Degree-of-Freedom (SDOF) oscillators. These SDOFs span the 1–10 Hz frequency band, encompassing low, medium, and high eigenfrequencies, with their accuracy forming an integral part of the model's evaluation. The results are promising, as AV measurements in an image format after being submitted to the trained model become free of additive noise. This with the aid of upscaling enables the possibility of deriving target eigenfrequencies without altering or deforming of them. Comparisons in various terms, both qualitative and quantitative, such as the mean magnitude-squared coherence, mean phase difference, and Signal-to-Noise Ratio (SNR), showed great performance.

Keywords: ambient vibration; convolutional neural networks; denoising; machine learning; structural health monitoring

1. Introduction

Structural Health Monitoring (SHM) has become an intriguing topic during the last decades and is applied to various fields of civil, mechanical, automotive, and aerospace engineering, among others. Estimating the health condition and understanding the unique characteristics of structures by assessing the measured physical parameters in real time represents a major objective of SHM. As a result, signal processing has become an essential and inseparable part of methodologies introduced via research related to SHM. The application of signal processing techniques to the structural damage identification procedure is classified into two types of approaches, namely, (i) time-domain and (ii) frequency-domain methods. Experimental studies have assessed the potential of signal processing techniques in the two aforementioned domains, aiming to enhance vibration-based structural damage detection subjected to environmental effects (earthquakes, wind, etc.).

Although multiple review studies have been published on vibration-based structural damage detection, there have been no studies on categorizing signal processing techniques based on feature extraction procedures that belong to time and frequency domains for SHM purposes. Recently, this was explored in the work by Zhang et al. [1]. Meanwhile, with the developments in the classical SHM approach, neural networks and big data analytics have paved the way for a new approach in the field of SHM. As Zinno et al. [2] showcased with their work, Artificial Intelligence (AI) could benefit SHM applications for bridge structures in several phases: construction, development, management, and maintenance. Moreover, buildings are aging, and deriving newer architectural trends while preserving building heritage is not only of high value but also a complex procedure requiring multi-criteria approaches to decision making to be implemented [3]. Therefore, AI-based methodologies specially tailored to assist the preservation of building heritage through SHM techniques have already been developed and are summarized in the work by Mishra [4]. Generally speaking, one of the most common implementations of deep learning methodologies in SHM applications relies on the convolutional neural network (CNN) architecture; a summary of these can be found in the recent work by Sony et al. [5]. As is mentioned in the conclusion of the aforementioned work, one of the targets of future research will be the development of the real-time implementation of CNN-based approaches in everyday SHM practice; the model presented in the current work was developed for real-world practice. CNN-based models have already been applied with great success in different scientific fields, e.g., the work by Xu et al. [6], which also inspired the authors to develop the model presented in the current study; more specifically, in the work by Xu et al. [6], a CNN and a Recurrent Neural Network (RNN) were combined in order to dynamically detect levels of ambient noise from speech gaps and remove them from audio signals without distorting the speech audio quality. Also, during the last decades, multiple Deep Learning models have found a successful place in the denoising task for various noisy images. An overview about this can be seen in Elad et al.'s [7] work and Izadi et al.'s [8] work. In a recent investigation, Damikoukas and Lagaros [9] explored the feasibility of utilizing an ML model as a robust tool to predict building earthquake responses, addressing the shortcomings of simplified models. The study advanced by integrating AV measurements and earthquake time-history data into a neural network framework. This innovative model presents a promising pathway to deepen our comprehension of structural behavior in the face of seismic events, thereby contributing to the advancement of earthquake resilience in building design and engineering. Other recent works that tried to take advantage of neural networks in structural engineering and seismic response estimation are those of Xiang et al. [10] and Demertzis et al. [11].

The motivation behind this research was to delve into the potential of Deep Learning in addressing denoising challenges associated with Micro-Electromechanical System (MEMS) digital sensors, specifically accelerometers. This category of sensors is notable for its cost-effectiveness compared to alternatives like force balance accelerometers. However, they often contend with elevated levels of electronic noise. Recognizing the ongoing strides in AI capabilities and their ubiquitous integration into various domains, the motivation behind this work was to leverage Deep Learning for denoising tasks, laying the groundwork for more advanced applications, such as the real-time denoising of acceleration time histories. MEMS digital sensors, while economically advantageous, can suffer from inherent electronic noise. This noise can compromise the accuracy of measurements, especially in dynamic applications like acceleration monitoring. By harnessing the power of Deep Learning, this research aims to effectively mitigate this noise and enhance the reliability of sensor outputs. The broader context lies in the rapid evolution of AI technologies, which are progressively becoming integral components of everyday applications. This research envisions the deployment of Neural Processing Unit (NPU) hardware at the measurement site, paving the way for the real-time denoising of acceleration time histories. This not only addresses current challenges but also anticipates future scenarios where advanced techniques are seamlessly integrated into the measurement process. The utilization of NPUs represents a strategic move toward decentralized processing, enabling on-site denoising without relying solely on external computational resources. As a result, this research endeavors to contribute to the practical implementation of AI-driven denoising techniques, fostering advancements in the field of sensor technology and real-time data processing.

The achievement of this study is that ambient vibration (AV) measurements are processed through a properly calibrated neural network (NN) in an image format, and the structural response is unveiled after removing the additive electrical noise from the AV recordings. To train the model, 1197 structural oscillators (models of Single Degree of Freedom (SDOF)) were developed, from which 10,773 numerically produced noisy signals were generated. These signals were converted into images in order to be fed to the NN-based model chosen for the purposes of this study. For validation purposes, the results were converted back to numerical values in order to assess the level of denoising, among other factors, in terms of frequency spectra between predicted and target signals. The proposed model is called MLDAR, which stands for Machine Learning-based Denoising of Ambient Response.

The main contributions of this paper are (i) the denoising of ambient vibration (AV) acceleration measurements (ii) by presuming the structural eigenfrequencies in domain spectra needed for further structural analysis and assessment, all of which are realized by (iii) exploiting the power of neural networks (NNs) and Deep Learning (DP) deploying a multi-convolutional and transposed convolutional network.

The remainder of this paper is organized as follows. Section 2 generally describes the characteristics of the structural response generated by ambient vibration, how research on SHM during the last several decades has been progressing, and how new technologies are finding their fit due to their versatility and cost efficiency. In Section 2, the structural parameters used for the creation of the dataset are also presented, along with the list of the assumptions implemented. Afterward, Section 3 introduces the Machine Learning-based Denoising of Ambient Response (MLDAR) model, the basic principles on which it was based on, and also all the chosen parameters of its structure. Thereafter, Section 4 showcases the validation results of the proposed MLDAR model, both qualitative and quantitative. The paper concludes with some final remarks given in Section 5 and insights into the authors' future work.

2. Structural Response Due to Ambient Vibration and SDOF Models

In this section, the characteristics of the structural response generated by means of ambient vibration is provided, together with the details of the models used to calibrate the neural network model developed for the purposes of this study.

2.1. Structural Response Generated by Ambient Vibration

Structures are permanently induced by various types of site excitations, which refer to either ground vibrations related to nature, e.g., earthquake genesis, or those generated by humans, like the vibrations generated during excavations, traffic, construction works, etc. Thus, structures continuously vibrate due to the above-mentioned causes, offering the possibility of monitoring and studying the structural response permanently. The structural response is of high value for structural engineers, as, among other tasks, they can derive the dynamic characteristics of the structure through the measurements they collect. Structural Health Monitoring (SHM) is a field where multiple sensors are deployed in order to enable engineers to monitor and assess the structural integrity, assist in deciding which interventions should be implemented, or even be alerted to an event, either at the precautionary or post-event state (early warning, etc.).

As nowadays, advancements in the manufacturing and technology of sensors and microcontrollers are huge, more and more sensing devices are released without any discounts on the level of their quality, shifting SHM research away from the traditional wired acceleration-sensing systems [12–15]. Meanwhile, attempts to broaden the use of SHM have been made, as more and more standards and design codes are being revised where SHM appears as an option or as an obligation for the engineers [16–19]. Therefore, there is an opportunity to ease any disadvantages of low-cost monitoring devices in terms of their implementation in civil engineering projects, bringing SHM into the mainstream of the structural engineering profession.

A common disadvantage of low-cost sensing devices is the higher level of noise that is introduced into the measurements. This noise is the additive distortion of the true-value signals. When it comes to high-magnitude motions, it is not an issue, as the real values can be acquired and processed easily. However, when we are referring to the ambient response of structures, the magnitude scale is really small, even smaller than today's low-cost accelerometers' level of noise. Therefore, in order to make use of noisy measurements in various algorithms and methodologies, time-domain dynamic quantities are usually skipped in favor of frequency-domain ones, such as eigenfrequencies. There are various data processing techniques that unveil information hidden by noise, such as the averaging of Fourier Spectra (e.g., [20]) and the recent work by the authors [21]. As said before, in the time domain, little can be done when measurements are already "noisy", and here comes the current work to fill the gap, taking advantage of the power of neural networks and their image detection capabilities.

2.2. *Models Used for Calibrating the NN*

Neural networks, in order to be designed and trained, need both parameters of a system: input(s) and output(s). From that point on, after we decide properly what architecture we want and correctly configure every parameter, they can learn all those complex relationships between the prediction and what we have as input data. In this work, the approach is straightforward, as the aim is to train the neural network to distinguish additive noise from a noisy signal, which includes an ambient vibration structural response, and remove it in order to return a "clean" ambient vibration structural response signal. In order to do this, we would ideally need the "clean" and "noisy" versions of lots of response signals for different kinds of sensors and buildings. As that is not something easily feasible, we decided to start building the network from the ground up, using numerical data that corresponds to a batch of assumptions made both for sensor specifications and for the building models themselves.

Therefore, all data/measurements used were numerically generated and computationally derived from Newmark numerical integration. The reason was that we wanted to cover a whole range of SDOF oscillators, with all the possible different parameters' combinations. In total, 1197 oscillators were used for the purposes of this study. The assumptions used to construct these oscillators were based on the model building that can be seen in Figure 1. The dimensions and other properties of the model building are specifically representative of the floor plan typically found in residential concrete buildings in Greece. This deliberate choice ensures that our study is grounded in a context reflective of prevalent architectural norms in the region, contributing to the relevance and applicability of our findings within the specified context. This model was constructed in the ADINA analysis software [22].

ADINA stands as a versatile software package renowned for its extensive capabilities in finite element analysis (FEA) and computational fluid dynamics (CFD). Tailored to address a diverse array of engineering challenges spanning multiple disciplines, ADINA excels in solving problems related to structural mechanics, heat transfer, fluid dynamics, electromagnetics, and multiphysics simulations. The acronym "ADINA" itself encapsulates its core functionality, representing "Automatic Dynamic Incremental Nonlinear Analysis". As a comprehensive tool, ADINA's strength lies in its ability to perform intricate simulations, offering engineers and researchers a robust platform for tackling complex problems in fields ranging from structural engineering to fluid dynamics. The software's broad applicability makes it an invaluable resource for professionals seeking accurate and efficient solutions across various domains of engineering. ADINA's prowess in dynamic, incremental, and nonlinear analyses underscores its suitability for simulating real-world scenarios, where the interactions between components and materials exhibit complex and dynamic behavior. By seamlessly integrating these capabilities, ADINA empowers users to gain deeper insights into the performance of their designs, aiding in the optimization of structures, processes, and systems. In essence, ADINA stands at the forefront of engineering simulation software, providing a sophisticated and adaptable suite of tools for addressing the multifaceted challenges inherent in the realm of finite element analysis and computational fluid dynamics. The number of assumptions and parameters used are shown in Table 1.

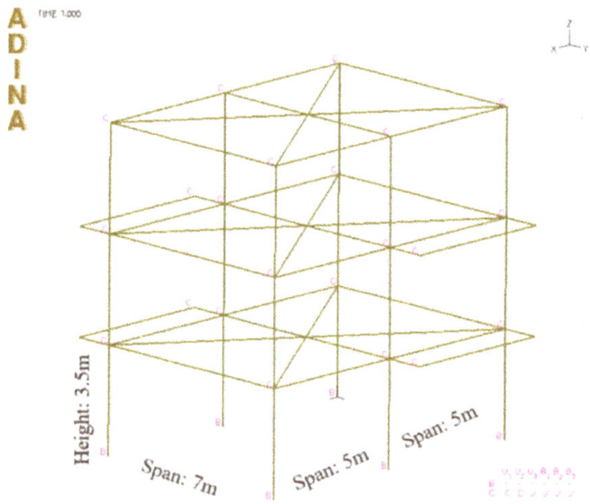

Figure 1. Typical building model in ADINA on which assumptions-table is referring to.

Table 1. Model parameters.

Geometry	
Plan	10.00 × 7.00 (m^2)
Stories	1 to 7
Story height	3.50 (m)
Slab thickness	0.25 (m)
Columns	0.50 × 0.50 (m^2)
Beams	0.40 × 0.70 (m^2)
Loads	
Dead	806.75 (kN)
Live	806.75 (kN)
Safety factor	1 [1]
Dynamic characteristics	
Mass (per story)	110.78 (tons)
Damping ratio ζ	5%
Eigenfrequency	1 to 10 Hz with step of 0.5
Material	
Reinforced concrete	
Bilinear material	Figure 2
Yield point	0.0105 m [2]
Post-yield stiffness	50% of geometric one [3]

[1] Assessment of existing conditions—real loads; [2] 0.003 drift × 3.50 m = 0.0105 m (HazusC3L—LowCode). More details can be found in Hazus®–MH 2.1 Technical Manual (see Paragraph 5.2.1 of [23]); [3] EC8–1 ($K_e = 0.5K_g$). After first yield, loading is still happening with K_{eff}, even for small forces than F_y.

Figure 2 depicts the bilinear capacity curve used for all structural members. For each structural element, we considered two distinct stiffness states: geometrical stiffness, denoted by K_g, and effective stiffness, represented by K_{eff}. The effective stiffness, K_{eff}, is assumed to be half of the geometrical stiffness (K_g) and arises from the degradation experienced by the member under deformation, particularly in conditions of higher loads. This dual representation of stiffness allows for a more nuanced understanding of the structural response. The geometrical stiffness, K_g, captures the inherent stiffness of the member in its ideal, undistorted state, while the effective stiffness, K_{eff}, accounts for the impact of deformations and degradation induced by higher loads. By acknowledging the dynamic interplay between these two stiffness states, the model provides a more accurate and comprehensive depiction of the member's behavior under varying conditions, facilitating a more realistic simulation of structural performance.

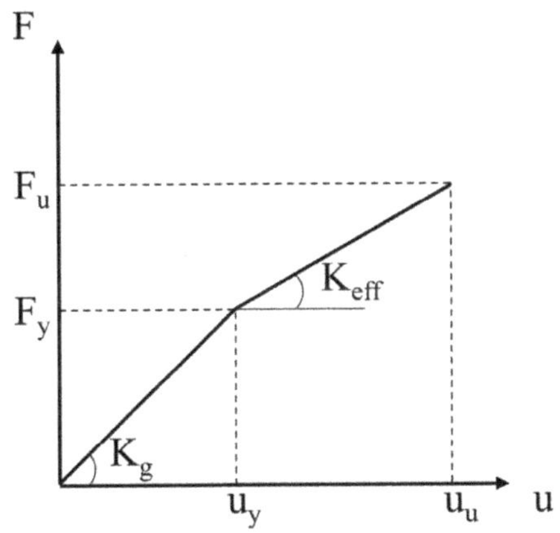

Figure 2. Bilinear capacity curve of all structural members.

The numerically created ambient acceleration responses of the building's models are noiseless, as they would be measured in an ideal experimental world, where measuring devices do not interfere in the slightest with measurable quantities. However, in real-world engineering applications, not only quantities are recorded by means of monitoring devices, as they are accompanied by various levels of noise, such as electronic noise. This family of noises is mathematically described as white noise, and statistically, it follows a normal distribution, with an average value of zero and a standard deviation related to the noise levels of the corresponding measuring device. The Signal-to-Noise Ratio (SNR), Noise Density (e.g., $\mu g/\sqrt{Hz}$), and others are terms that usually describe the levels of noise in the recorded signals.

The final signals are the result of the sum of the aforementioned signals. The sampling rate is 100 Hz, and the duration of the artificial recordings is 60 s. Each building model consists of 1 to 7 floors, with a mass ranging between 80% and 120% of the aforementioned typical values (see Table 1) and with an eigenfrequency ranging between 1 and 10 Hz with a step of 0.5 Hz. Therefore, 1197 models of single-stage oscillators were derived. For each of these models, there are three signal windows of the theoretically no-(additional)-noise response and three electronic and non-electronic noise signal windows, which, superpositioned, lead to nine combinations of final response signals (pure oscillator response + electronic/other noise). Thus, 10,773 artificial signals were generated, which were converted into images and were the input of the machine learning model. They were created in the Matlab environment (release R2021b [24]). Of the generated signals, 75% were used for training (training set), while the other 25% were the validation sample (validation set).

3. Machine Learning-Based Models: Architecture and Calibration

During the last decade, due to the advances achieved in computer technology, machine learning has become very popular, having been applied with great success in different scientific areas, like autonomous vehicles, visual recognition, news aggregation and fake news detection, robotics, natural language processing, vocal Artificial Intelligence (AI), etc. Convolutional neural networks (CNNs) represent a class of artificial neural networks (ANNs) most commonly used for analyzing images. What makes them unique is that the network learns to optimize the filters (or kernels) through automated learning, whereas in traditional algorithms, these filters are hand-engineered. This independence from prior knowledge and human intervention in feature extraction is a major advantage. CNNs have applications in image and video recognition, recommender systems, image classification, image segmentation, medical image analysis, natural language processing, brain–computer interfaces, and financial time series.

3.1. Convolutional Neural Networks

Convolutional networks (e.g., LeCun et al. [25]), also known as convolutional neural networks, or CNNs, are a specialized kind of neural network for processing data that has a known grid-like topology. Examples include time-series data, which can be thought of as a 1-D grid taking samples at regular time intervals, and image data, which can be thought of as a 2-D grid of pixels. Convolutional networks have been tremendously successful in practical applications. The name "convolutional neural network" indicates that the network employs a mathematical operation called convolution.

$$s(t) = \int x(a) \cdot w(t-a)da \tag{1}$$

where $x(t)$ is the raw signal measurement at time t, $w(a)$ is a weighted average that gives more weight to recent measurements, a denotes the age of a measurement, and $s(t)$ is the smoothed estimate of the $x(t)$ measurement.

Convolution is also denoted as follows:

$$s(t) = (x \cdot w)(t) \tag{2}$$

In convolutional network terminology, the first argument (the function x) to the convolution is often referred to as the input, and the second argument (the function w) is the kernel. The output is sometimes referred to as the feature map (Figure 3). In our case, as also in many others, the convolution taking place is two-dimensional, and time is discrete. Therefore, its mathematical notation, called convolution without flipping, which is equivalent to cross-correlation, is defined as follows:

$$s(i,j) = (K \cdot I)(i,j) = \sum_m \sum_n I(i+m, j+n) \cdot K(m,n) \tag{3}$$

where I is a two-dimensional array of data (e.g., an image), and K is a two-dimensional kernel; both I and K are discrete values.

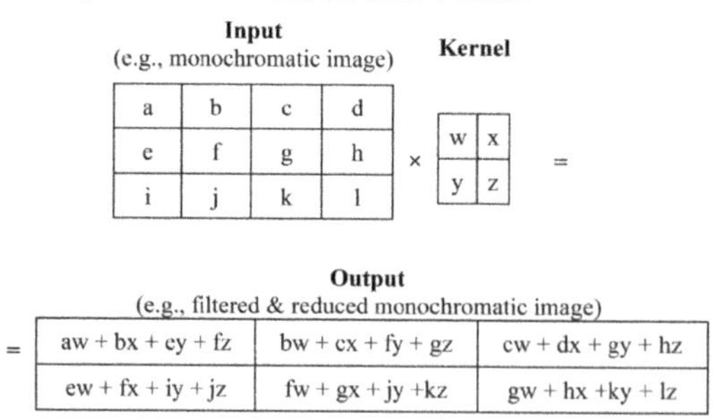

Figure 3. A 2D CNN channel.

3.2. The Autoencoders

An autoencoder is a neural network that is trained to attempt to copy its input to its output (Figure 4). Internally, it has a hidden layer **h** that describes a code used to represent the input. The network may be viewed as consisting of two parts: an encoder function $\mathbf{h} = f(\mathbf{x})$ and a decoder that produces a reconstruction $\mathbf{r} = g(\mathbf{h})$. If an autoencoder succeeds in simply learning to set $g(f(\mathbf{x})) = \mathbf{x}$ everywhere, then it is not especially useful. Instead, autoencoders are designed to be unable to learn to copy perfectly. Usually, they are restricted in ways that allow them to copy only approximately and to copy only input that resembles the training data. Because the model is forced to prioritize which aspects of the input should be copied, it often learns useful properties of the data. Denoising autoencoders must undo corrupted/noisy input measurements rather than simply copying them as they are.

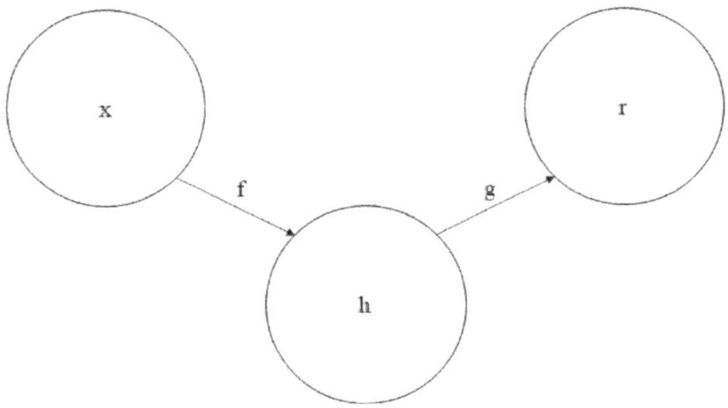

Figure 4. The general structure of an autoencoder, mapping an input x to an output (called reconstruction) r through an internal representation or code h. The autoencoder has two components: the encoder f (mapping x to h) and the decoder g (mapping h to r).

3.3. MLDAR: A Machine Learning-Based Model for Denoising the Ambient Structural Response

In this study, a machine learning-based model is presented that is able to denoise ambient-response recordings collected using instruments with specifications equivalent to those of a low-cost monitoring device (noise density: 22.5 $\mu g/\sqrt{Hz}$, MEMS type) used in recent SHM studies [21,26–28]. The proposed model is labeled as MLDAR, which stands for Machine Learning-based Denoising of Ambient Response, and is presented graphically in Figure 5. In particular, MLDAR refers to a denoising autoencoder type of neural network whose purpose is to reproduce its inputs' time histories of ambient structural responses free of noise.

Autoencoders serve as powerful tools for denoising tasks, leveraging their inherent capacity to extract meaningful features from input data while effectively filtering out noise. Comprising both an encoder and a decoder, autoencoders are designed to learn the mapping of noisy input to a lower-dimensional representation, facilitating the subsequent reconstruction of the clean input. The training process involves optimizing the model to minimize the disparity between the reconstructed and pristine inputs, a mechanism that inherently eradicates noise. The denoising prowess of autoencoders arises from their fundamental need to discern between signals and noise. This process highlights the autoencoder's capability to emphasize salient features within the data, rendering it particularly adept at tasks that demand precision amidst ambient noise. A pertinent application of this capability can be observed in the context of Micro-Electromechanical Systems (MEMs), where autoencoders prove invaluable in enhancing the accuracy of signal extraction in the presence of inherent electronic noise. In essence, autoencoders stand out as versatile tools for noise reduction, excelling in tasks that require the meticulous separation of signals from noise. Their ability to learn intricate patterns in data and prioritize essential features positions them as valuable assets in applications ranging from MEMs to various domains where precision amidst ambient noise is paramount.

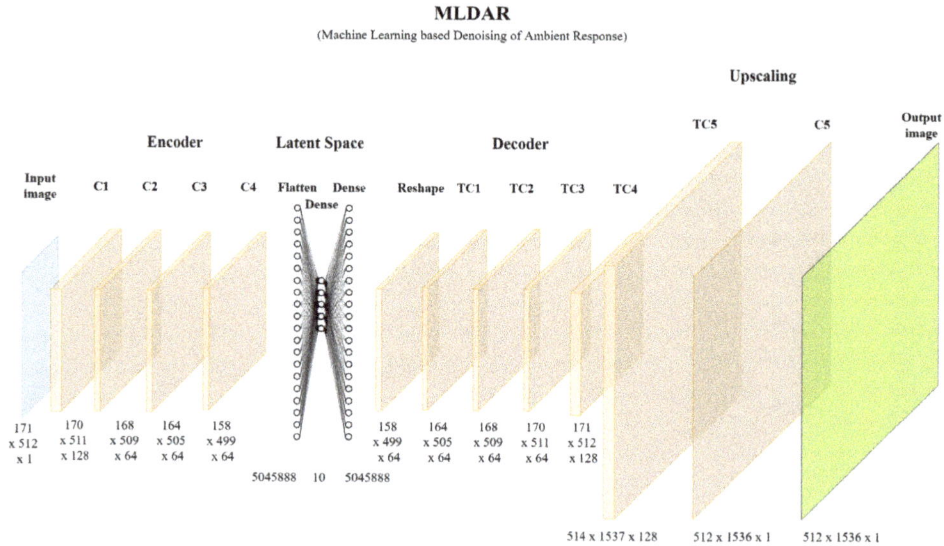

Figure 5. The MLDAR neural network model.

The proposed NN model consists of the **encoder**, the **latent space** representation, the **decoder** part, and last but not least, the **upscaling** part. The input to the proposed model is the noisy ambient response, which can be viewed as a 2D image of size $T \times F$ with two channels, where T represents the time length of the time-history signal, and F denotes the amplitude of acceleration (in g_s) at the given time. On the other hand, the output of the model is the corresponding ambient response, clear of any additive noise (e.g., electronic). Its form is also a 2D image with the same size, $T \times F$, as the input.

The signal that precedes the generation of the image takes the form of a time history of acceleration, organized with a consistent timestep. Essentially, it can be envisioned as a one-dimensional matrix encompassing acceleration values, supplemented by a concealed column that incorporates the temporal dimension. This representation is designed to highlight the temporal evolution of the acceleration data, accentuating their dynamic nature in the context of image creation. By structuring the information in this manner, we aim to provide a comprehensive understanding of how acceleration changes over time, offering valuable insights into the intricate process of image generation. The signals throughout this work are not presented as graphs but as images. Therefore, they do not have X- and Y-axes noted in the images themselves. However, they are implied. The X-axis corresponds to samples, and the Y-axis represents acceleration values.

These images are normalized in the range of 0 to 1, a well-known practice for artificial neural networks, depending also on the activation functions adopted. For the needs of the study, a computing system equipped with a standalone NVIDIA Titan RTX graphics card with 24 GB VRAM was used to implement the training part. Relying on the VRAM capabilities, the I/O images' dimensions were chosen as 512×171 pixels for the input and as 1536×512 pixels for the output. Moreover, in addition to the noise elimination process, an image upscaling process took place with the use of some further levels in the neural network due to input-image compression due to hardware restrictions. The 510 width pixels of input compressed images practically mean that the sampling rate was reduced from 100 Hz to 8.53 Hz. The purpose of the upscaling part is to add the required details to the output signals in order to enable further signal processing techniques, such as the Fast Fourier Transform (FFT), and restore as much of the lost frequency spectrum as possible. After implementing the upscaling process, a sampling rate of 25.6 Hz is restored, meaning that a bandwidth of 12.8 Hz is retained in our signals.

In order to convert images from acceleration values, the following limits were used:

- Input images (noisy response): minimum value of −0.000442 g, maximum value of 0.000437 g;
- Output images (no-noise response): minimum value of −0.000074 g, maximum value of 0.000072 g.

The characteristics of the proposed MLDAR model (Figure 5) are the following: the images introduced as input arguments are discretized with 512 × 171 pixels; therefore, the input format is [171, 512, 1], where 171 refers to the number of pixels along the height of each image, 512 is the number of pixels along the width of each image, and 1 is the value of the monochrome channel (grayscale) of the image. The image is then passed through a 2D convolutional encoder in order to extract the features of major importance. The 2D **encoder** Table 2 consists of four layers of 2D convolution (Conv2D), and *ReLu* was selected as the proper activation function. Every Conv2D layer is followed by Batch Normalization and Dropout (with a frequency of 25%) layers in order to avoid overfitting and so that the network is able to generalize with better accuracy. After implementing the encoding part, two dense layers are used for the **latent space** representation in order to achieve the desirable compression of the feature values. Feature values were unable to be directly interpreted at the image-pixel input layer; however, it is now possible by means of encoding the **latent space** representation in a reduced multidimensional space. *ReLu* is also used as the proper activation function after the small dense layer. Having reshaped the last dense layer to a two-dimensional one, the **decoder** Table 3 is used in order to distance from the feature maps of latent space and return to the composition of the monochrome image. The decoder consists of four levels of Conv2DTranspose, and each of them is complemented by the *ReLu* activation function. Finally, the **upscaling** section Table 3 is implemented, which enlarges the image to the desired output size of 1536 × 512 pixels. Upscaling is implemented at two levels, the Conv2DTranspose and the Conv2D one. The first level relies on a *ReLu* activation function, while the last one relies on a **Sigmoid** suppression function, since the output refers to an image with a color value ranging between 0 to 1. Regarding the activation functions, the Sigmoid activation function is characterized by an output range confined within [0,1]. This property proves particularly advantageous when dealing with grayscale images, as their pixel values typically span the range of 0 to 255 (or normalized to 0–1). Sigmoid effectively scales and squashes the output to a probability-like range, aligning well with the characteristics of normalized image data. Conversely, the Rectified Linear Unit (ReLU) activation function introduces crucial nonlinearity to the model, enabling it to discern and learn intricate patterns and representations within the data. Empirically, ReLU has demonstrated robust performance, notably facilitating accelerated convergence during training. In the context of convolutional autoencoders, ReLU is commonly employed in both the encoding and decoding layers. This choice is deliberate, aiming to capture and preserve essential features in the data. The rectification operation inherent in ReLU aids in the learning of hierarchical and spatial features, enhancing the model's capacity to extract meaningful information from the input.

The loss function chosen corresponds to the mean absolute error between the true label and the prediction (Equation (4)):

$$loss = 1/n \sum_{t=1}^{n} \mid e_t \mid = 1/n \sum_{t=1}^{n} \mid y_{pred} - y_{true} \mid \tag{4}$$

The batch size was chosen to be six sets of images (input–output). The training took place in 80 epochs. The total number of training parameters was 106,375,947. The Adam optimizer was used on this network with a learning step equal to 0.0001. At the last epoch, the training error was 0.0045, while the prediction error was 0.0053 (in Figure 6, see the history of the training process). The exploration of various model structures, hyperparameters, and training techniques is a crucial aspect of Deep Learning. This iterative process entails systematic testing, evaluation, and subsequent refinement guided by performance metrics. In the initial phase of parameter calibration, our emphasis primarily rested on qualitative comparisons of the results. As the calibration progressed, a transition was made to a more quantitative assessment approach, allowing for a comprehensive analysis of the model's performance and efficacy. This sequential evolution in our methodology ensures a thorough and balanced evaluation of the experimented variables throughout the experimentation process.

Table 2. Architecture of MLDAR—Part I. "C" indicates a convolutional layer.

	Encoder			
	C1 [1]	**C2 [1]**	**C3 [1]**	**C4 [1]**
Filters:	128	64	64	64
Kernel size:	(2,2)	(2,2)	(2,2)	(2,2)
Dilation:	(1,1)	(2,2)	(4,4)	(6,6)
Stride:	(1,1)	(1,1)	(1,1)	(1,1)
Padding:	Valid	Valid	Valid	Valid

[1] Followed by Batch Normalization, ReLu, and Dropout layers.

Table 3. Architecture of MLDAR—Part II. 'C' indicates a convolutional layer, and "TC" indicates a transposed convolutional layer.

	Decoder				Upscaling	
	TC1 [1]	**TC2 [1]**	**TC3 [1]**	**TC4 [1]**	**TC5 [1]**	**C5 [2]**
Filters:	64	64	64	128	128	1
Kernel size:	(2,2)	(2,2)	(2,2)	(2,2)	(4,4)	(3,2)
Dilation:	(6,6)	(4,4)	(2,2)	(1,1)	(1,1)	(1,1)
Stride:	(1,1)	(1,1)	(1,1)	(1,1)	(3,3)	(1,1)
Padding:	Valid	Valid	Valid	Valid	Valid	Valid

[1] Followed by ReLu; [2] followed by Sigmoid.

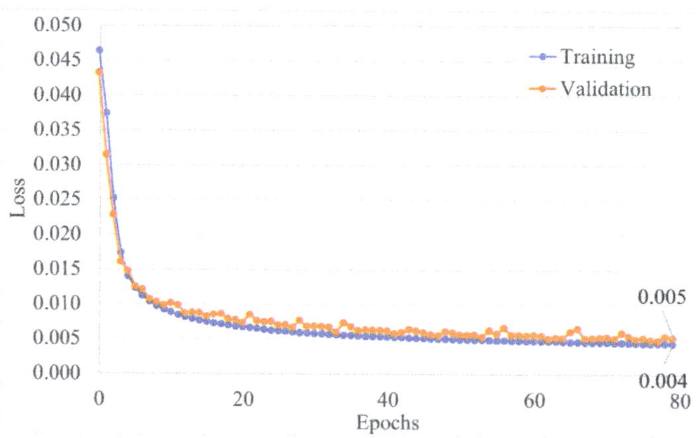

Figure 6. Training and validation loss function values over 80 epochs.

4. Frequency Spectrum Comparison: Qualitative and Quantitative Results

In this part of the study, some of the results obtained in the framework of the investigation are presented. In particular, low-, medium-, and high-frequency signals are randomly selected, aiming to present the efficiency of the proposed MLDAR model for a spectrum of frequency values. Apart from the comparison of the results obtained in terms of images, the frequency content is also compared, the outcome that was the primary goal of the present work.

4.1. Qualitative Comparison: Sample of Low-Frequency Signals

Many mid- to high-rise buildings, bridges, and other flexible structures are usually those showing low-frequency spectral responses, i.e., usually in the range of 1.0 to 3.0 Hz. The dynamic characteristics of these ground-induced structures correspond to a lower acceleration amplitude and a higher displacement amplitude. However, the damping part takes a longer time to diminish the responses to this type of structure. To examine the efficiency of the MLDAR model for such low-frequency cases, a sample denoted as 4_1 was randomly chosen; this sample refers to structural model #4, while the #1 time-window is used to represent the noise. The characteristics of the specific structural model #4 include a frequency value equal to 1 Hz, a one-story structure, and a mass equal to 95% of the reference mass provided in Table 1. The noisy signal can be seen in Figure 7, and the non-noisy one can be found in Figure 8, while the predicted one obtained through the MLDAR model is shown in Figure 9.

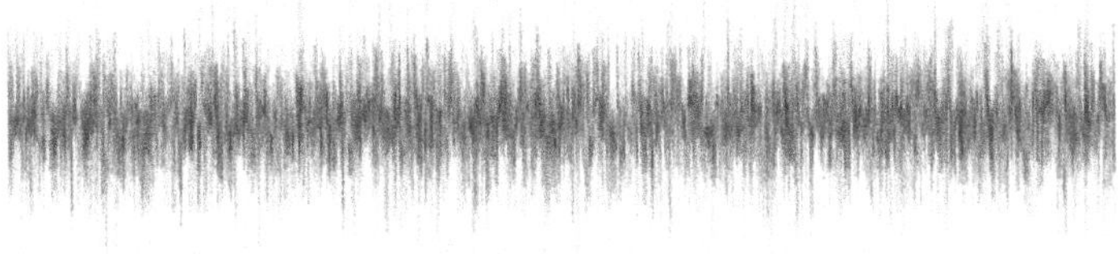

Figure 7. The noisy low-frequency signal (signal 4_1).

Figure 8. The low-frequency signal without noise [Target] (signal 4_1).

Figure 9. The denoised low-frequency signal through the MLDAR model [Prediction] (signal 4_1).

In order to make the difference in the magnitude scale (noisy vs. clean signal) more clearly visible, an amplitude comparison between the signal with and without additive noise is provided in Figure 10. This specific sample of noisy signal has values that range in 7.75×10^{-4} mg with a standard deviation of 1.02×10^{-4} mg. On the other hand, its counterpart, the target non-noisy one, has values that range in 7.27×10^{-5} mg with a standard deviation of 9.69×10^{-6} mg. As can be seen, the range difference is almost 10 times. To this point, it should be pointed out again that the noisy signal is sampled at 100 Hz, and the non-noisy one, the product and target of the proposed NN, is sampled at 25.6 Hz. The reduction in the sampling rate consequently reduces the level of the total noise. Therefore, calculating the average power of a signal at different sampling rates results in different values. In this case, signal 4_1 in the original 100 Hz form has an average power of −79.79 dB, that at 25.6 Hz has an average of −85.81 dB, and the target signal sampled at 25.6 Hz has an average power of −98.80 dB.

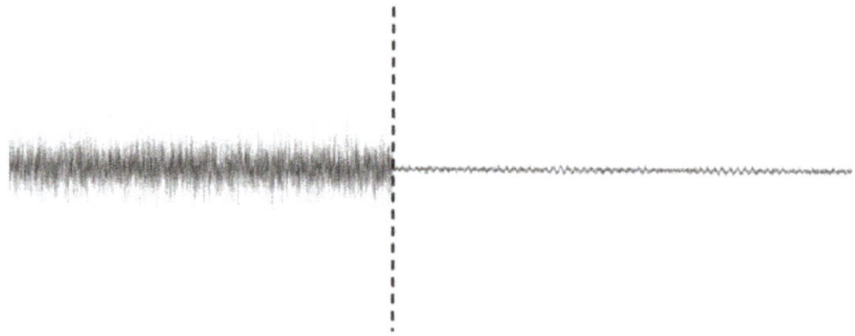

Figure 10. Low-frequency case: scale comparison between noisy signal and non-noisy (cleaned–predicted).

Regarding the frequency content of the signals generated by the specific model case, a comparison at the frequency-domain level can be seen in Figure 11. The blue-colored line depicts the noisy signal sampled at 100 Hz, and that was used in the .png format as input in the NN. Orange and violet correspond to the target and predicted non-noisy signals, which show the response of model #4 under ambient vibrations. As can be seen, the 1 Hz frequency is indistinguishable in the ambient vibration measurements due to noise. Even after the averaging of the Fast Fourier Transforms (FTTs) with a 1 min time-window, the 1 Hz frequency cannot be identified. On the other hand, the implementation of the MLDAR model managed to remove the noise from the signal at an acceptable level, where the extraction of the 1 Hz frequency is feasible. As can be seen, the whole frequency band is much lower in the predicted and target signals, showing that the baseline high-level electronic noise has been removed. A comparison in time-history terms can be seen in Figure 12, where differences are colored in green and magenta, and matching regions are in gray.

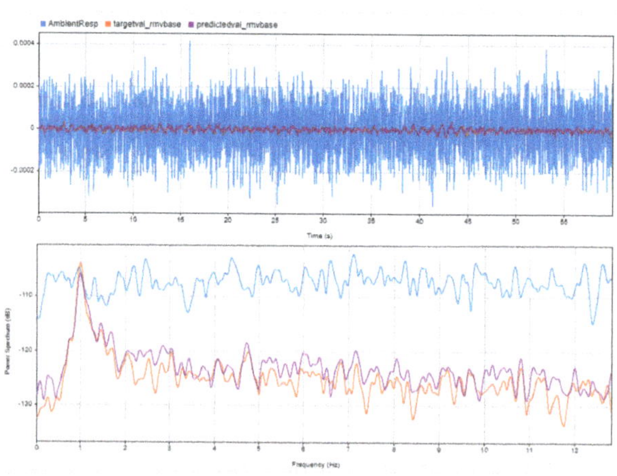

Figure 11. Low-frequency case: comparison of frequency content between response to ambient noise (blue), target signal (orange), and predicted signal (violet)—frequency of 1.0 Hz.

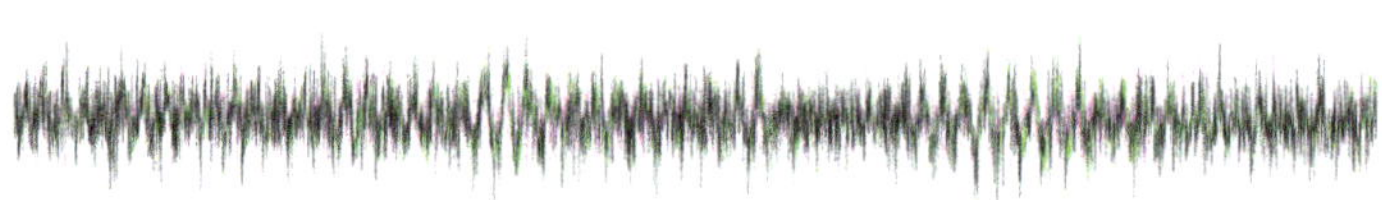

Figure 12. Low-frequency case: comparison of frequency content between target signal (orange) and prediction (light blue)—frequency of 1.0 Hz.

4.2. Qualitative Comparison: Sample of Medium-Frequency Signal

Many mid-rise buildings, including typical concrete buildings designed based on older building codes, are usually those showing medium-frequency spectral responses, i.e., usually in the range of 3.0 to 6.0 Hz. In order to further examine the efficiency of the MLDAR model, a sample denoted by 631_4 was randomly chosen, belonging to the medium-frequency cases. This sample refers to structural model #631, while the #4 time-window was used to represent the noise. The characteristics of structural model #631 include a frequency value equal to 6 Hz, and similar to the previous case study, it is a one-story structure, and the mass is equal to 80% of the reference mass provided in Table 1. The noisy signal for this specific case can be seen in Figure 13, and the non-noisy one can be found in Figure 14, while the predicted one obtained through the MLDAR model is depicted in Figure 15.

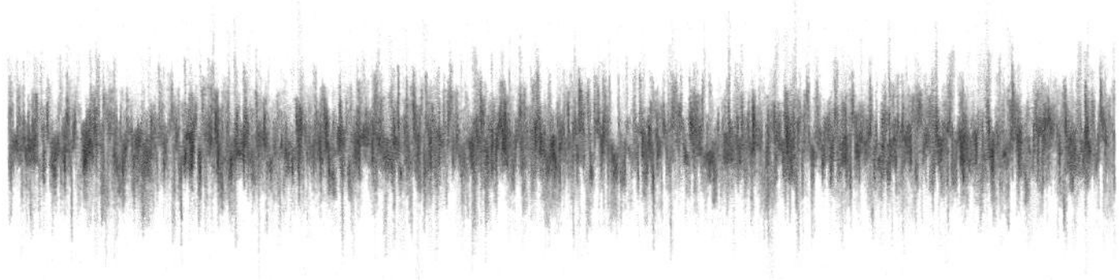

Figure 13. The noisy medium-frequency signal (signal 631_4).

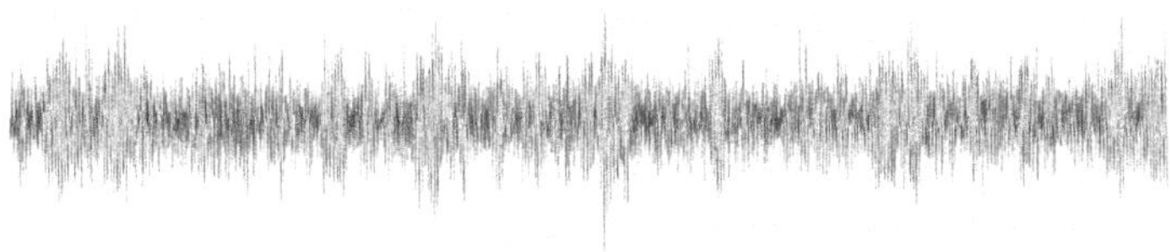

Figure 14. The medium-frequency signal without noise [Target] (signal 631_4).

Similar to the low-frequency case, and in order to make the difference in the magnitude scale (noisy vs. clean signal) more clearly visible, an amplitude comparison between the signal with and without additive noise is provided in Figure 16. This specific sample of a noisy signal has values that range in 7.78×10^{-4} mg with a standard deviation of 1.03×10^{-4} mg. On the other hand, its counterpart, the target non-noisy one, has values that range in 7.20×10^{-5} mg with a standard deviation of 1.08×10^{-5} mg. As can be seen, the range difference is almost 10 times. For this case, signal 631_4 at the 100 Hz sampling rate has an average power of -79.76 dB, that at 25.6 Hz has an average of -85.69 dB, and the target signal sampled at 25.6 Hz has an average power of -95.53 dB.

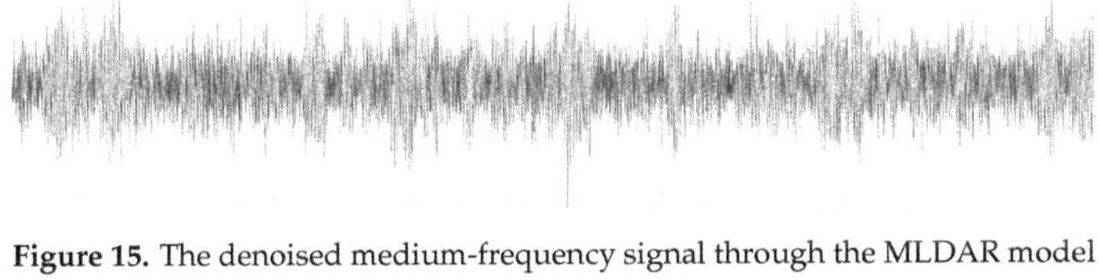

Figure 15. The denoised medium-frequency signal through the MLDAR model [Prediction] (signal 631_4).

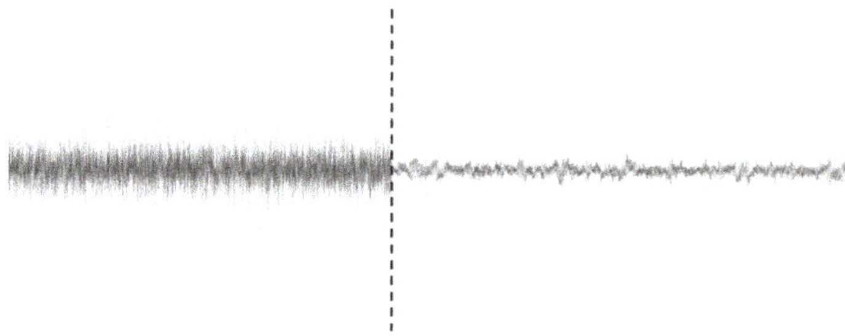

Figure 16. Medium-frequency case: scale comparison between noisy signal and non-noisy (cleaned–predicted).

Regarding the frequency content of the signals generated by the specific model case, a comparison at the frequency-domain level can be seen in Figure 17. The blue-colored line depicts the noisy signal sampled at 100 Hz, and that was used in the .png format as input in the NN. Orange and violet correspond to target and predicted non-noisy signals, which shows the response of model #631 under ambient vibrations. As can be seen, the 6 Hz frequency is indistinguishable in the ambient vibration measurements due to noise. Even after the averaging of the Fast Fourier Transforms (FTTs) with a 1 min time-window, the 6 Hz frequency cannot be identified, as its intensity is much smaller than that of the noise band. On the other hand, the implementation of the MLDAR model managed to remove the noise from the signal at an acceptable level, where the extraction of the 6 Hz frequency is feasible. As can be seen, the whole frequency band is much lower in the predicted and target signals, showing that the baseline high-level electronic noise has been removed. A comparison in time-history terms can be seen in Figure 18, where differences are colored in green and magenta, and matching regions are in gray.

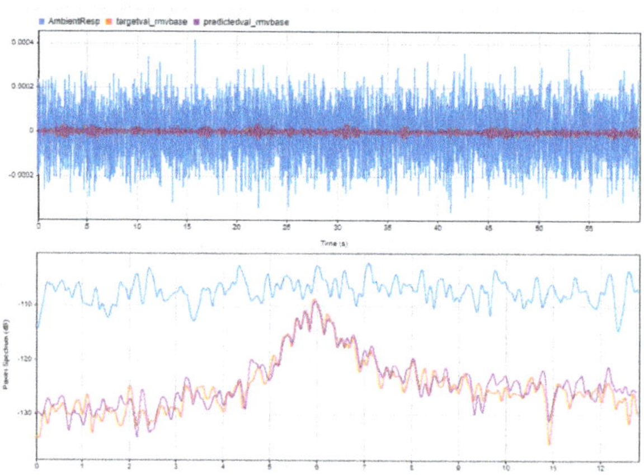

Figure 17. Medium-frequency case: comparison of frequency content between response to ambient noise (blue), target signal (orange), and predicted signal (violet)—frequency of 6.0 Hz.

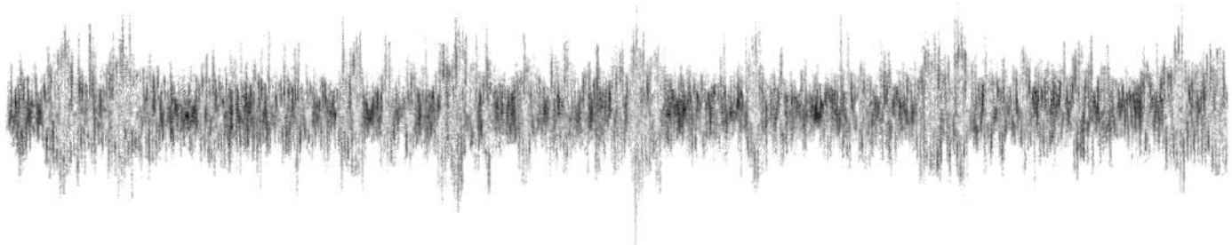

Figure 18. Medium-frequency case: Comparison of frequency content between target and predicted non-electronically noisy signals. Differences are highlighted by orange and magenta colors, and matching regions are in gray—frequency of 6.0 Hz.

4.3. Qualitative Comparison: Sample of High-Frequency Signal

Low-rise buildings such as those designed and built based on modern design codes or older masonry building structures are usually those showing high-frequency spectral responses, i.e., usually equal to 6.0 Hz or higher. The dynamic characteristics of these ground-induced structures correspond to a higher acceleration amplitude, which, however, is damped at a higher rate of response attenuation. To further examine the efficiency of the proposed MLDAR model for higher-frequency cases, a sample denoted by 1194_7 was randomly chosen. This sample refers to structural model #1194, while the #7 time-window was used to represent the noise. The characteristics of structural model #1194 include a frequency value equal to 10 Hz, a seven-story structure, and a mass equal to 105% of the reference mass provided in Table 1. The noisy signal can be seen in Figure 19, and the non-noisy one can be found in Figure 20, while the predicted one obtained through the MLDAR model is shown in Figure 21.

Similar to the other two cases, for the high-frequency case as well, in order to make the difference in the the magnitude scale (noisy vs. clean signal) more clearly visible, an amplitude comparison between the signal with and without additive noise is provided in Figure 22. This specific sample of noisy signal has values that range in 8.01×10^{-4} mg with a standard deviation of 1.04×10^{-4} mg. On the other hand, its counterpart, the target non-noisy one, has values that range in 6.94×10^{-5} mg with a standard deviation of 1.17×10^{-5}. As can be seen again, the range difference is almost 10 times. In this case, signal 631_4 at a 100 Hz sampling rate has an average power of −79.70 dB, that at 25.6 Hz has an average of −85.56 dB, and the target signal sampled at 25.6 Hz has an average power of −92.51 dB.

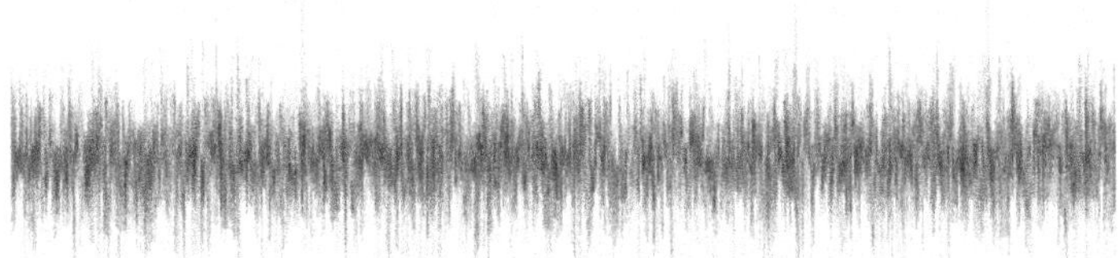

Figure 19. The noisy high-frequency signal (signal 1194_7).

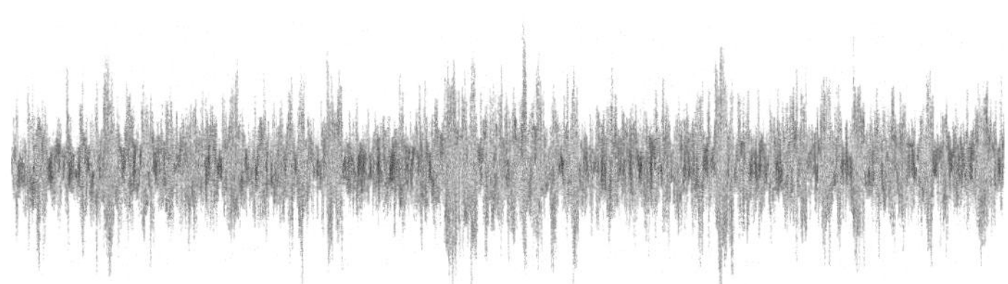

Figure 20. The high-frequency signal without noise [Target] (signal 1194_7).

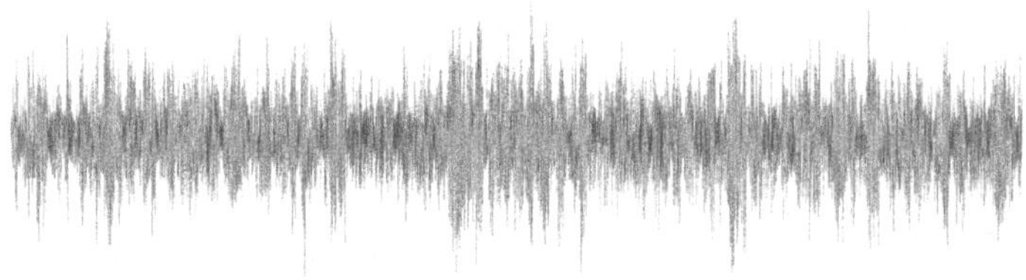

Figure 21. The denoised high-frequency signal through the MLDAR model [Prediction] (signal 1194_7).

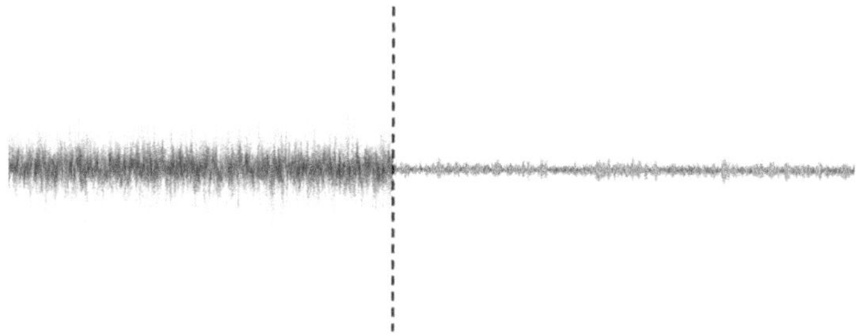

Figure 22. High-frequency case: scale comparison between noisy signal and non-noisy (cleaned–predicted).

Regarding the frequency content of the signals generated by the specific model case, a comparison at the frequency-domain level can be seen in Figure 23. The blue-colored line depicts the noisy signal sampled at 100 Hz, and that was used in the .png format as input in the NN. Orange and violet correspond to target and predicted non-noisy signals, which shows the response of model #1194 under ambient vibrations. As can be seen, the 10 Hz frequency is indistinguishable in the ambient vibration measurements due to noise. Even after the averaging of the Fast Fourier Transforms (FTTs) with a 1 min time-window, the frequency of interest cannot be identified, as its intensity is much smaller than that of the noise band. On the other hand, the implementation of the MLDAR model managed to remove the noise from the signal at an acceptable level, where the extraction of the target frequency is feasible. As can be seen, the whole frequency band is much lower in the predicted and target signals, showing that the baseline high-level electronic noise has been removed. A comparison in time-history terms can be seen in Figure 24, where differences are colored in green and magenta, and matching regions are in gray.

4.4. Quantitative Results: Comparing Frequency Spectra of Prediction and Target for the Whole Dataset through Magnitude-Squared Coherence

Although the denoising process that relies on a Deep Learning model (MLDAR model) concerns acceleration time-history recordings and operates on their image formatting, the main objective of the original problem is the frequency extraction from ambient vibration measurements; thus, the comparison should also be performed on the efficiency of the MLDAR model in extracting eigenfrequencies from the denoised ambient response signals. The first step is to convert the MLDAR image-output dataset into numerical time histories, and then the frequency spectra need to be extracted through the Fast Fourier Transform (FFT) algorithm.

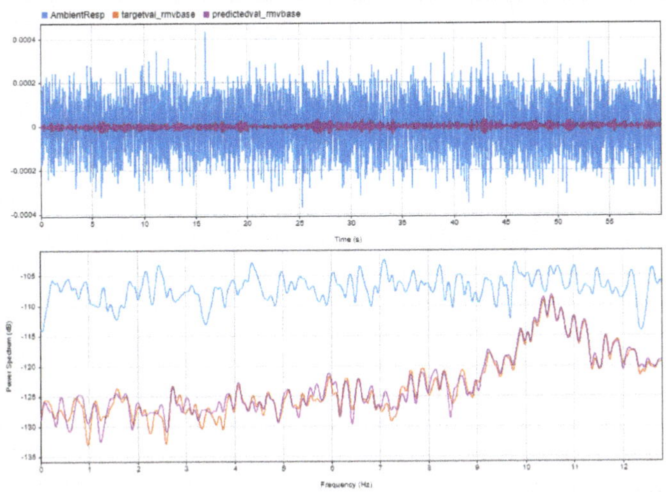

Figure 23. High-frequency case: comparison of frequency content between response to ambient noise (blue), target signal (orange), and predicted signal (violet)—frequency of 10.0 Hz.

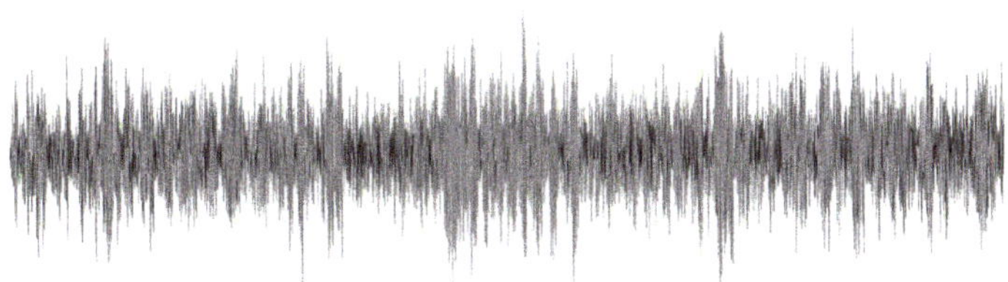

Figure 24. High-frequency case: Comparison between target and predicted non-electronically noisy signals. Differences are highlighted by orange and magenta colors, and matching regions are in gray—frequency of 10.0 Hz.

To compare the efficiency of the MLDAR method, a comparison in frequency terms is also performed between predicted and target (original and non-noisy) signals for the validation dataset (i.e., 2700 signals). The validation dataset contains around 25% of the total generated 10,773 signals. Specifically, the magnitude-squared coherence values $C_{xy}(f)$ (Equation (5)) [29–31] are calculated for a specified frequency range between predicted and target signals. Then, the mean value of $C_{xy}(f)$ for each sample pair of signals is derived. This frequency range depends on the SDOF frequency of interest of each signal and is determined as follows: $f_{SDOF} \pm 0.1$ (Hz). As seen in Figure 25, the minimum mean $C_{xy}(f)$ value for the whole validation dataset is 91%. The average value of the mean $C_{xy}(f)$ for the whole validation dataset was calculated at 0.98, with a standard deviation of 0.01. The standard deviation of $C_{xy}(f)$ for each sample varies, as shown in Figure 26.

$$C_{xy}(f) = \frac{|P_{xy}(f)|^2}{P_{xx}(f)P_{yy}(f)}. \tag{5}$$

where $C_{xy}(f)$ is the magnitude-squared coherence of the x and y signals, $P_{xx}(f)$ and $P_{yy}(f)$ are the power spectral densities of the two signals, and $P_{xy}(f)$ is the cross-power spectral density of the two signals. $C_{xy}(f)$ is between 0 and 1.

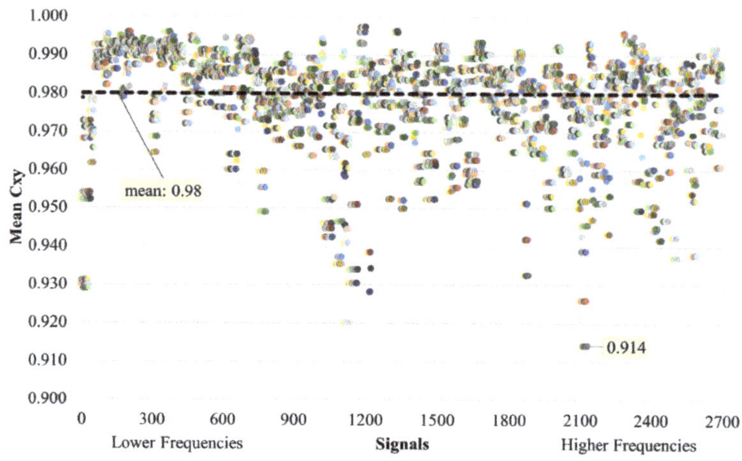

Figure 25. Mean magnitude-squared coherence $C_{xy}(f)$ for whole validation dataset between target and predicted signals.

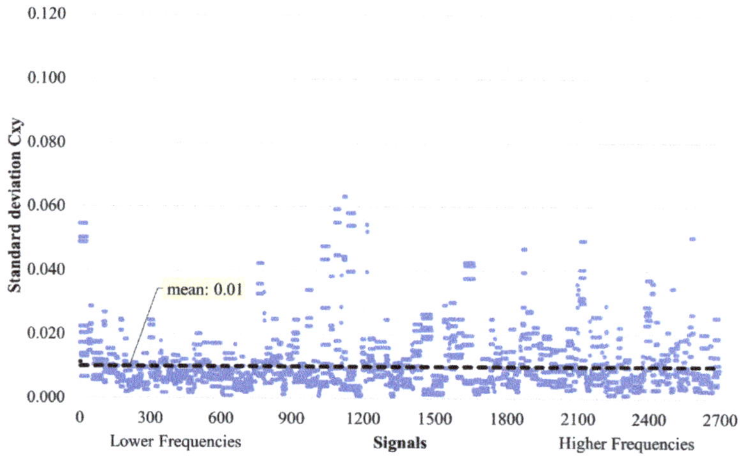

Figure 26. Standard deviation of magnitude-squared coherence $C_{xy}(f)$ for whole validation dataset between target and predicted signals.

For purposes of completeness, the phase difference ($\delta\phi$) between the target and predicted signals in the validation dataset is also presented based on the already-calculated $P_{xy}(f)$ values, i.e., the cross-power spectral density of the two signals. Similarly, for the same frequency range of each signal, the average of the difference in phase is calculated, and the trend for all signals is summarized in Figure 27. The average value of the mean ($\delta\phi$) for the whole dataset of signals is calculated as 0.20 degrees, with a standard deviation of 3.245. The standard deviation of $\delta\phi$ for each sample varies, as shown in Figure 28.

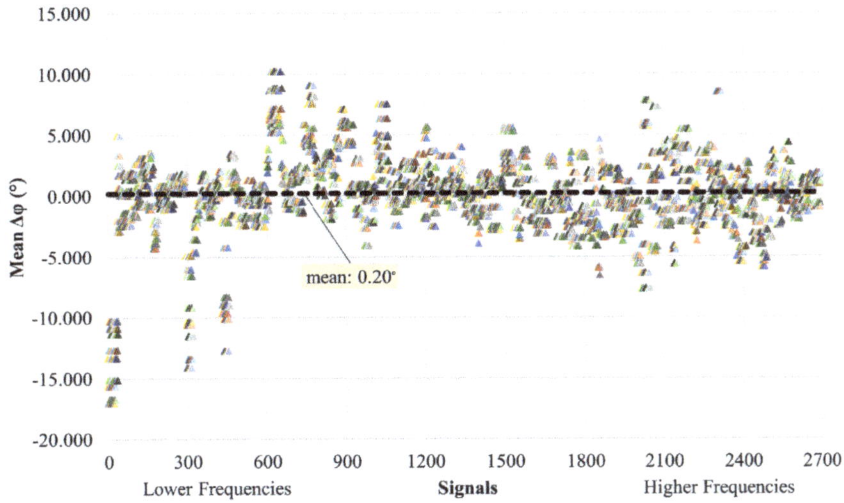

Figure 27. Mean phase difference ($\delta\phi$) for whole validation dataset between target and predicted signals.

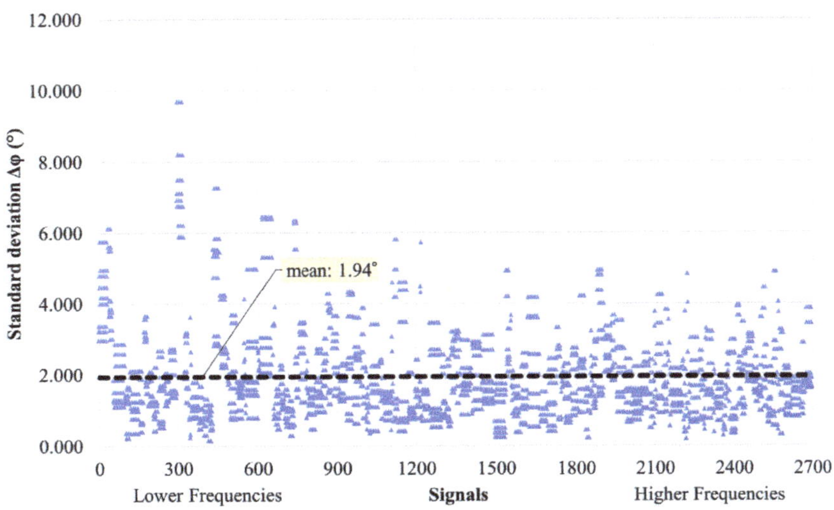

Figure 28. Standard deviation of phase difference ($\delta\phi$) for whole validation dataset between target and predicted signals.

Distinctive colors have been introduced in Figures 25 and 27, with the purpose of highlighting the individuality of each graph point, emphasizing their uniqueness within the dataset. The decision to employ varied colors is strategic, as using a uniform color for all points could result in a visually overwhelming amalgamation, resembling a continuous painted area due to the sheer volume of points within the limited horizontal space. By assigning unique colors to each point, we ensure that the values remain discrete and discernible, preventing the potential visual confusion that might arise from a homogeneous color scheme.

4.5. Quantitative Results: Evaluating Denoising Performance through SNR Levels

Finally, an additional index, the Signal-to-Noise Ratio (SNR) (Equation (6)), was employed to assess the denoising performance. The SNR is a commonly used metric in the fields of science and engineering to compare the level of a desired signal to the level of background noise. It is typically calculated as the ratio of the signal power to the noise power and is often expressed in decibels. An SNR ratio greater than 1:1 (exceeding 0 dB) indicates that the signal strength exceeds that of the noise, thereby signifying a favorable outcome.

$$SNR = \frac{P_{clean}}{P_{noisy}} \tag{6}$$

The SNR was calculated by taking the average power, denoted by P. The SNR was computed between the noisy and non-noisy signals, specifically between the target and predicted signals. This calculation was performed for the entire validation set, and the corresponding results are presented in Figure 29. SNR#1 refers to the ratio between the target signal and the noisy signal, while SNR#2 represents the ratio between the predicted signal and the noisy signal. The mean value of SNR#1 is found to be −8.40 decibels (dB), indicating a relatively low signal strength compared to the background noise. The median value for SNR#1 is −8.16 dB, suggesting a similar trend in the central tendency of the data. Similarly, the mean of SNR#2 is −8.44 dB, implying a comparable signal–noise relationship for the predicted signal. The median value for SNR#2 is −8.29 dB, reinforcing the observations made from the mean value. These SNR measurements provide quantitative information regarding the relationship between the target or predicted signals and the accompanying noise. The negative dB values indicate that the noise level tends to overshadow the signal strength, highlighting the need for further improvement in denoising techniques to enhance signal clarity.

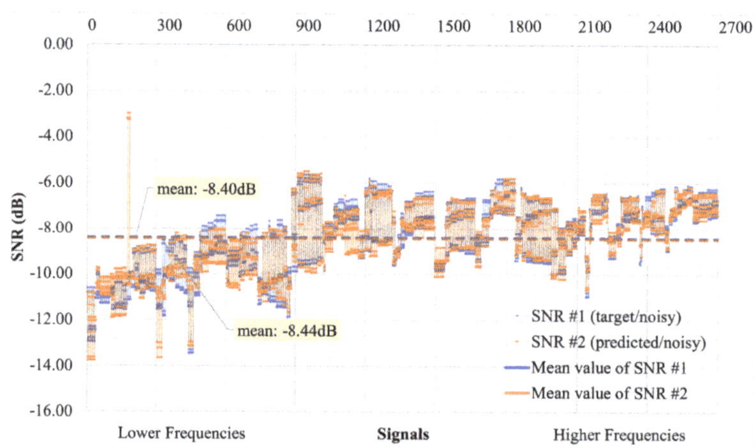

Figure 29. Signal-to-Noise Ratio (SNR) for whole validation dataset between target and predicted signals.

5. Conclusions

This research proposes the use of a neural network model as a denoiser for ambient vibration measurements, with the primary objective of removing noise while preserving the essential information necessary for subsequent signal manipulation. In this study, signal manipulation refers to the extraction of the dynamic characteristics, such as eigenfrequencies and eigenmodes, of Single-Degree-of-Freedom (SDOF) building models.

In pursuit of denoising various signals, challenges arise after employing a digital sensor or converting an analog sensor output into digital form, leaving limited options for noise reduction. One established approach involves the application of Digital Signal Processing (DSP) techniques, such as digital filtering, moving average filters, and moving median filters. These methods aim to effectively mitigate noise interference. An alternative strategy entails implementing averaging techniques in the frequency domain following the Fourier Transform. This approach proves particularly beneficial in addressing noise issues by exploiting the frequency characteristics of the signals. In the context of analog sensors, the quest for enhanced noise reduction requires a multifaceted approach. This includes the adoption of superior digitization hardware, exemplified by high-quality Analog-to-Digital Converters (ADCs). Furthermore, improvements in shielding for cables and the integration of advanced decoupling transistors contribute significantly to fortifying the system against unwanted noise. Nevertheless, in the specific context of this study, a distinctive choice was made to utilize a Deep Learning architecture for denoising purposes. This decision reflects a departure from traditional techniques, indicating an exploration into the innovative realm of machine learning for noise reduction. By opting for Deep Learning, this study endeavors to leverage the model's ability to discern complex patterns and extract relevant features, potentially offering a more sophisticated and adaptive solution to the challenges posed by signal noise. This strategic shift aligns with the evolving landscape of signal processing, embracing the promising capabilities of advanced machine learning architectures for effective denoising in diverse sensor applications.

To train and validate the denoiser model, a dataset of response signals was artificially generated based on existing accelerometer noise specifications. Both qualitative and quantitative evaluations demonstrate that the proposed MLDAR model effectively eliminates almost all types of additive noise, including electronic and non-electronic sources, from theoretically noise-free ambient response signals. Despite the significant difference in scale between noisy and noise-free signals, as evident in Figures 10, 16, and 22, the MLDAR model consistently succeeds in removing the noise from the signals. It is worth noting that the MLDAR model produces output signals with a resolution of 12.8 Hz, which is sufficient for most common building structures and civil engineering infrastructures. This limitation is due to the smaller resolution of the input signals compared to the output signals, potentially attributed to hardware capabilities. Nonetheless, the MLDAR model's upscaling capabilities open avenues for future work, enabling its utilization in versatile and lightweight applications, such as web applications and IoT devices. This makes it possible to combine the denoising model with earthquake building seismic assessment tools and methodologies.

Moreover, the quantitative results of the MLDAR model, as presented in the last paragraph of the numerical investigation section, exhibit its promising performance. The validation dataset shows that the worst performance achieved was a 91% accuracy for only a few cases, while the average score reached 98% (refer to Figure 25). This confirms the successful accomplishment of the primary goal of this study, which is extracting the eigenfrequencies of SDOF building models from noisy signals, a task previously challenging without extending the sampling time or employing statistical signal manipulation techniques [21].

The model was trained using a dataset composed of artificially generated ambient response signals designed to replicate the noise specifications of a MEMS-type accelerometer, specifically the ADXL355 model. These signals were superimposed on the ambient responses of Multi-Degree-of-Freedom (MDOF) building models. The outcomes showcased the efficacy of the MLDAR model in effectively eliminating additive noise from ostensibly noise-free ambient signals. Notably, the model demonstrated this capability despite the substantial scale difference between noisy and non-noisy signals. A constraint observed is the resolution of the output signal images, which restricts the signal's useful bandwidth to 12.8 Hz, given a sampling rate of 25.6 Hz. Nevertheless, this resolution proves adequate for the majority of building structures and applications in civil engineering. Quantitative assessments affirm the model's promising performance, with high magnitude-squared coherence scores averaging at 98%, coupled with minimal phase differences of 0.20°. An impressive outcome of the study is the MLDAR model's capacity to extract fundamental eigenfrequencies from MDOF building models even in the presence of noisy signals, achieving the primary objective without the necessity of extending the sampling time or employing statistical signal manipulation. To further enhance the capabilities of the MLDAR model in future endeavors, several avenues can be explored. Expanded Training Data: Consider retraining the model by incorporating additional data from diverse sensors and ambient vibration field measurements. Real-Time Implementation: Integrate the trained model into a microcontroller or single-board microcomputer/barebone equipped with AI capabilities. This would facilitate the real-time denoising of measurements. Notably, Tensorflow, even in its Tensorflow Lite version, is compatible with a range of low-cost devices (USD 50–150), such as Arduino Nano 33 BLE Sense, Espressif ESP32, Raspberry Pi 4, NVIDIA® Jetson Nano™, and Coral Dev Board. Versatility Improvement: Explore the possibility of enhancing the model's versatility in signal denoising tasks, including time-history signals of various natures. This expansion could broaden the applicability of the MLDAR model across a wider spectrum of scenarios.

Author Contributions: Conceptualization, S.D. and N.D.L.; methodology, S.D. and N.D.L.; software, S.D.; validation, S.D. and N.D.L.; investigation, S.D. and N.D.L.; resources, N.D.L.; data curation, S.D. and N.D.L.; writing—original draft preparation, S.D.; writing—review and editing, S.D. and N.D.L.; visualization, S.D.; supervision, N.D.L.; project administration N.D.L.; funding acquisition, N.D.L. All authors have read and agreed to the published version of the manuscript.

Funding: This research has been financed by the IMSFARE project "Advanced Information Modelling for SAFER structures against manmade hazards" (Project Number: 00356).

Institutional Review Board Statement: Not applicable.

Informed Consent Statement: Not applicable.

Data Availability Statement: Data are contained within the article.

Acknowledgments: It is acknowledged that the research was supported by the Hellenic Foundation for Research and Innovation (H.F.R.I.) under the "2nd Call for H.F.R.I. Research Projects to support Post-Doctoral Researchers", IMSFARE project "Advanced Information Modelling for SAFER structures against manmade hazards" (Project Number: 00356).

Conflicts of Interest: The authors declare no conflicts of interest.

Abbreviations

The following abbreviations are used in this manuscript:

AI	Artificial Intelligence
ANNs	Artificial neural networks
AV	Ambient vibration
CNN	Convolutional neural network
ML	Machine learning
MLDAR	Machine Learning-based Denoising of Ambient Response
MEMs	Micro-Electromechanical Systems
NN	Neural network
NPUs	Neural Processing Units
RNN	Recurrent Neural Network
SDOF	Single Degree of Freedom
SHM	Structural Health Monitoring
SNR	Signal-to-Noise Ratio

References

1. Zhang, C.; Mousavi, A.A.; Masri, S.F.; Gholipour, G.; Yan, K.; Li, X. Vibration feature extraction using signal processing techniques for structural health monitoring: A review. *Mech. Syst. Signal Process.* **2022**, *177*, 109175. [CrossRef]
2. Zinno, R.; Shaffiee Haghshenas, S.; Guido, G.; Vitale, A. Artificial Intelligence and Structural Health Monitoring of Bridges: A Review of the State-of-the-Art. *IEEE Access* **2022**, *10*, 88058–88078. [CrossRef]

3. Nadkarni, R.R.; Puthuvayi, B. A comprehensive literature review of Multi-Criteria Decision Making methods in heritage buildings. *J. Build. Eng.* **2020**, *32*, 101814. [CrossRef]
4. Mishra, M. Machine learning techniques for structural health monitoring of heritage buildings: A state-of-the-art review and case studies. *J. Cult. Herit.* **2021**, *47*, 227–245. [CrossRef]
5. Sony, S.; Dunphy, K.; Sadhu, A.; Capretz, M. A systematic review of convolutional neural network-based structural condition assessment techniques. *Eng. Struct.* **2021**, *226*, 111347. [CrossRef]
6. Xu, R.; Wu, R.; Ishiwaka, Y.; Vondrick, C.; Zheng, C. Listening to Sounds of Silence for Speech Denoising. *arXiv* **2020**, arXiv:2010.12013.
7. Elad, M.; Kawar, B.; Vaksman, G. Image Denoising: The Deep Learning Revolution and Beyond—A Survey Paper. *SIAM J. Imaging Sci.* **2023**, *16*, 1594–1654. [CrossRef]
8. Izadi, S.; Sutton, D.; Hamarneh, G. Image denoising in the deep learning era. *Artif. Intell. Rev.* **2023**, *56*, 5929–5974. [CrossRef]
9. Damikoukas, S.; Lagaros, N.D. MLPER: A Machine Learning-Based Prediction Model for Building Earthquake Response Using Ambient Vibration Measurements. *Appl. Sci.* **2023**, *13*, 10622. [CrossRef]
10. Xiang, P.; Zhang, P.; Zhao, H.; Shao, Z.; Jiang, L. Seismic response prediction of a train-bridge coupled system based on a LSTM neural network. *Mech. Based Des. Struct. Mach.* **2023**, 1–23. [CrossRef]
11. Demertzis, K.; Kostinakis, K.; Morfidis, K.; Iliadis, L. An interpretable machine learning method for the prediction of R/C buildings' seismic response. *J. Build. Eng.* **2023**, *63*, 105493. [CrossRef]
12. Kohler, M.D.; Hao, S.; Mishra, N.; Govinda, R.; Nigbor, R.L. *ShakeNet: A Portable Wireless Sensor Network for Instrumenting Large Civil Structures*; US Geological Survey: Reston, VA, USA, 2015.
13. Sabato, A.; Feng, M.Q.; Fukuda, Y.; Carní, D.L.; Fortino, G. A Novel Wireless Accelerometer Board for Measuring Low-Frequency and Low-Amplitude Structural Vibration. *IEEE Sens. J.* **2016**, *16*, 2942–2949. [CrossRef]
14. Sabato, A.; Niezrecki, C.; Fortino, G. Wireless MEMS-Based Accelerometer Sensor Boards for Structural Vibration Monitoring: A Review. *IEEE Sens. J.* **2017**, *17*, 226–235. [CrossRef]
15. Zhu, L.; Fu, Y.; Chow, R.; Spencer, B.F.; Park, J.W.; Mechitov, K. Development of a High-Sensitivity Wireless Accelerometer for Structural Health Monitoring. *Sensors* **2018**, *18*, 262. [CrossRef] [PubMed]
16. Zhou, L.; Yan, G.; Wang, L.; Ou, J. Review of Benchmark Studies and Guidelines for Structural Health Monitoring. *Adv. Struct. Eng.* **2013**, *16*, 1187–1206. [CrossRef]
17. Yang, Y.; Li, Q.; Yan, B. Specifications and applications of the technical code for monitoring of building and bridge structures in China. *Adv. Mech. Eng.* **2017**, *9*, 1687814016684272. [CrossRef]
18. Moreu, F.; Li, X.; Li, S.; Zhang, D. Technical Specifications of Structural Health Monitoring for Highway Bridges: New Chinese Structural Health Monitoring Code. *Front. Built Environ.* **2018**, *4*, 10. [CrossRef]
19. Yang, Y.; Zhang, Y.; Tan, X. Review on Vibration-Based Structural Health Monitoring Techniques and Technical Codes. *Symmetry* **2021**, *13*, 1998. [CrossRef]
20. Dunand, F.; Gueguen, P.; Bard, P.Y.; Rodgers, J. Comparison of the dynamic parameters extracted from weak, moderate and strong building motion. In Proceedings of the 1st European Conference of Earthquake Engineering and Seismology, Geneva, Switzerland, 3–8 September 2006.
21. Damikoukas, S.; Chatzieleftheriou, S.; Lagaros, N.D. First Level Pre- and Post-Earthquake Building Seismic Assessment Protocol Based on Dynamic Characteristics Extracted In Situ. *Infrastructures* **2022**, *7*, 115. [CrossRef]
22. *ADINA*, version: 9.2.1; ADINA R&D Inc.: Watertown, MA, USA, 2016. Available online: http://www.adina.com/ (accessed on 19 September 2022).
23. FEMA. *Hazus—MH 2.1: Technical Manual*; Department of Homeland Security, Federal Emergency Management Agency, Mitigation Division: Washington, DC, USA, 2013.
24. *MATLAB*, version 9.11.0 (R2021b); The MathWorks Inc.: Natick, MA, USA, 2021.
25. LeCun, Y.; Boser, B.; Denker, J.S.; Henderson, D.; Howard, R.E.; Hubbard, W.; Jackel, L.D. Backpropagation Applied to Handwritten Zip Code Recognition. *Neural Comput.* **1989**, *1*, 541–551. [CrossRef]
26. Shabani, A.; Feyzabadi, M.; Kioumarsi, M. Model updating of a masonry tower based on operational modal analysis: The role of soil-structure interaction. *Case Stud. Constr. Mater.* **2022**, *16*, e00957. [CrossRef]
27. Unquake. Unquake Accelerograph-Accelerometer Specifications. Available online: https://unquake.co (accessed on 19 September 2022).
28. Chatzieleftheriou, S.; Damikoukas, S. Optimal Sensor Installation to Extract the Mode Shapes of a Building—Rotational DOFs, Torsional Modes and Spurious Modes Detection. Technical Report. Unquake (Structures & Sensors P.C.). 2021. Available online: https://www.researchgate.net/institution/Unquake/post/6130bc2ca1abfe50c1559a26_Download_White_Paper_Optimal_sensor_installation_to_extract_the_mode_shapes_of_a_building-Rotational_DOFs_torsional_modes_and_spurious_modes_detection (accessed on 19 September 2022).
29. Welch, P. The use of fast Fourier transform for the estimation of power spectra: A method based on time averaging over short, modified periodograms. *IEEE Trans. Audio Electroacoust.* **1967**, *15*, 70–73. [CrossRef]

30. Rabiner, L.R.; Gold, B. *Theory and Application of Digital Signal Processing*; Prentice-Hall: Englewood Cliffs, NJ, USA, 1975.
31. Kay, S. Spectral Estimation. In *Advanced Topics in Signal Processing*; Prentice-Hall, Inc.: Englewood Cliffs, NJ, USA, 1987; pp. 58–122.

computation

MDPI

Article

Numerical Covariance Evaluation for Linear Structures Subject to Non-Stationary Random Inputs

M. Domaneschi [1], R. Cucuzza [1,*], L. Sardone [1], S. Londoño Lopez [1], M. Movahedi [2] and G. C. Marano [1]

[1] Department of Structural, Geotechnical and Building Engineering, Politecnico di Torino, Corso Duca degli Abruzzi, 24, 10129 Turin, Italy; marco.domaneschi@polito.it (M.D.); laura.sardone@poliba.it (L.S.); santiago.londono@polito.it (S.L.L.); giuseppe.marano@polito.it (G.C.M.)

[2] Department of Structural and Geotechnical Engineering, Széchenyi István University, Győr, Hungary

* Correspondence: raffaele.cucuzza@polito.it

Abstract: Random vibration analysis is a mathematical tool that offers great advantages in predicting the mechanical response of structural systems subjected to external dynamic loads whose nature is intrinsically stochastic, as in cases of sea waves, wind pressure, and vibrations due to road asperity. Using random vibration analysis is possible, when the input is properly modeled as a stochastic process, to derive pieces of information about the structural response with a high quality (if compared with other tools), especially in terms of reliability prevision. Moreover, the random vibration approach is quite complex in cases of non-linearity cases, as well as for non-stationary inputs, as in cases of seismic events. For non-stationary inputs, the assessment of second-order spectral moments requires resolving the Lyapunov matrix differential equation. In this research, a numerical procedure is proposed, providing an expression of response in the state-space that, to our best knowledge, has not yet been presented in the literature, by using a formal justification in accordance with earthquake input modeled as a modulated white noise with evolutive parameters. The computational efforts are reduced by considering the symmetry feature of the covariance matrix. The adopted approach is applied to analyze a multi-story building, aiming to determine the reliability related to the maximum inter-story displacement surpassing a specified acceptable threshold. The building is presumed to experience seismic input characterized by a non-stationary process in both amplitude and frequency, utilizing a general Kanai–Tajimi earthquake input stationary model. The adopted case study is modeled in the form of a multi-degree-of-freedom plane shear frame system.

Keywords: non-stationary random process; covariance analysis; Lyapunov equation; dynamic response and reliability

1. Introduction

Taking into account how structures behave in a random vibration setting is a prevalent method used to assess actual response scenarios [1]. This applies to various contexts, e.g., aircraft, vibrating machinery, and buildings subjected to marine or wind vibrations. These engineering scenarios involve examining how structures respond to dynamic and nondeterministic actions, and random dynamic analysis proves to be the most effective mathematical tool for this purpose [2]. This decision arises from the inherent randomness in inputs, a well-documented aspect in the field of random vibration theory (as evidenced by [3–6]). Approaching the problem using these methodologies enables acquiring relevant and reliable information about the structural response, typically unattainable through deterministic methods. The strength of the random vibration approach lies in its high-quality information, including the quantification of structural integrity, which is significant in probabilistic safety assessments. Failure is generally described as the initial moment when a structural crisis begins, tying it to the first instance where one or more measurements of structural response exceed a safe range. This usually involves assessing structural response indicators like displacements, stresses, buckling loads, or natural frequencies.

The random vibration problem for linear mechanical systems subject to Gaussian processes input is posed in stationary environmental conditions as the solution of the so-called Lyapunov matrix equation [7,8] to obtain the response covariance that defines completely the statistics of the system. Challenges in solving the Lyapunov equation have constrained the size of the meshes that could be employed. The Lyapunov equation (i.e., Lyapunov matrix equation, $AR + RA^T + B = 0$, A system matrix, R covariance matrix, and B input matrix, see Section 2) is typically achieved using algorithms like Bartels–Steward or Hessenberg–Schur. These methods require the Schur factorization of system matrix A. Various software tools for scientific computing, such as Matlab and Python, employ adapted versions of these algorithms, delivering satisfactory results for small dense matrices A and B. This involves O (N^3) floating-point operations and O (N^2) memory [9]. Approaches designed for large system dimensions have been developed, for instance, the Krylov subspace methods [9–11] or the matrix sign function decomposition with Newton's iterative method.

In non-stationary cases, under the same assumptions (linear system and Gaussian input) the time covariance approach is more complex as the Lyapunov matrix covariance becomes a differential one (i.e., Lyapunov differential matrix equation, $\dot{R} = AR + RA^T + B = 0$, see Section 3) whose numerical solution is sometimes more complex, and there are not standard tools to be implemented, differently from the stationary case.

The main objective of this research is to introduce a numerical technique for evaluating response covariance in the time domain for linear structures subjected to non-stationary stochastic loads. This method is tailored for a generic scenario where the input involves a non-stationary modulated filtered white noise process, capable of simulating various real physical loads like earthquakes [12–15].

To achieve a versatile non-stationary approach applicable across different contexts, the structural response is assessed using a covariance approach, as understanding the evolving covariance matrix in the space state is crucial for evaluating reliability, particularly in terms of initial failure events.

To address this, a time-step integration algorithm is proposed, employing the Euler implicit method to solve the differential Lyapunov matrix equation. The outcome is a sequential algorithm that requires the numerical solution of a stationary Lyapunov matrix equation at each time step, a task achievable through standard numerical tools. This method is implemented to reduce computational expenses and has been specifically applied to a multi-story building represented by a shear frame structure, examining dynamic responses under seismic base motions and assessing the reliability concerning initial threshold crossings.

2. Linear Elastic MDoF Subject to Non-Stationary Random Vibration

Many instances of real-life structural issues revolve around configurations that match a linear viscoelastic system of lumped masses. These systems face either steady or fluctuating forces. This study considers both scenarios to evaluate how structural responses vary statistically, employing the covariance approach [13,14].

This method presents notable benefits, especially in dynamic conditions, where inputs are simulated as white noises. These can be filtered to better match real dynamic occurrences. By solving the equations of the dynamic equilibrium system, this technique gauges the structural response of a deterministic second-order linear mechanical system composed of lumped masses when subjected to probabilistic dynamic input.

$$\mathbf{M}\overline{\ddot{X}_s}(t) + \mathbf{C}\overline{\dot{X}_s}(t) + \mathbf{K}\overline{X_s}(t) = \mathbf{G}_s\overline{f}(t) \tag{1}$$

where $\overline{X_s}$, $\overline{\dot{X}_s}$ and $\overline{\ddot{X}_s}$ are the structural displacement, velocity, and acceleration process vectors. **M, C** and **K** are the mass, viscous, and stiffness symmetric matrices. While the mass matrix is always positive definite, the damping and stiffness matrices are positive semi-definite. The vector

$$\overline{f}(t)^T = [f_1(t), f_2(t), \ldots, f_n(t)] \tag{2}$$

accumulates n stochastic excitations applied to the structure, while **G**s represents an m x n matrix linking the excitation components of the forcing vector to the structural degrees of freedom. When the elements of the system excitations vector are stationary white noises, the first- and second-order statistical moments remain unchanged over time.

$$\left\langle f_i^{ST}(t) \right\rangle = \mu_{f_i} \tag{3}$$

$$\left\langle f_i^{ST}(t_1) f_j^{ST}(t_2) \right\rangle = [\mathbf{J}]_{ij}\delta(t_2 - t_1) = \left[\mathbf{R}_{ff}^{ST}(t_2, t_2)\right]_{ij} \tag{4}$$

Moreover, if $f_i^{ST}(t)$ are Gaussian excitations, then the responses and their time derivatives constitute a Markov vector in the dimension phase state.

The matrix, related to a vector satisfying the shot noise properties (usually denoted as a shot noise vector), has diagonal elements equal to the autocovariance intensity of each force and extra-diagonal elements representing the level of correlation between two generic different forces, so that can vary from if and f_j are completely correlated, to zero, if and are completely un-correlated.

Then, in the case of complete un-correlated forcing loads, the matrix is replaced by the simpler diagonal matrix of components, where the elements are the input power spectral density of each entry.

A commonly used method involves expressing a non-stationary input through an intensity modulation of a stationary process, often referred to as uniform modulation. This method assumes that the intensity of the process alters over time according to a deterministic function $\varphi(t)$, while the spectral contents remain constant. Consequently, in the case of time modulation, a stationary forcing process vector is substituted by the following non-stationary vector:

$$\overline{f}^{NS}(t) = \left[\varphi_1(t) f_1^{ST}(t), \varphi_2(t) f_3^{ST}(t), \ldots, \varphi_n(t) f_n^{ST}(t)\right] \tag{5}$$

with the stochastic characterization

$$\left\langle f_i^{NS}(t) \right\rangle = \varphi_i(t)\mu_{f_i} \tag{6}$$

$$\left\langle f_i^{NS}(t_1) f_j^{NS}(t_2) \right\rangle = \varphi_i(t_1)\varphi_j(t_2)[\mathbf{J}]_{ij}\delta(t_2 - t_1) = \left[\mathbf{R}_{ff}^{NS}(t_1, t_2)\right]_{i,j} \tag{7}$$

and, in case of un-correlated excitations, the covariance matrix $\mathbf{R}_{ff}^{NS}(t_2, t_2)$ is diagonal

$$\mathbf{R}_{ff}^{NS}(t_1, t_2) = \left\langle f_i^{NS}(t_1) f_j^{NS}(t_2) \right\rangle = \begin{cases} 2\pi S_0^i \varphi_i(t_1)\varphi_i(t_2)\delta(t_2 - t_1) \text{ if } i = j \\ 0 \text{ if } i \neq j \end{cases} \tag{8}$$

Pre Filters Technique

In time-domain stochastic analysis, two primary methods are typically employed. The first, applicable when the input's autocorrelation function is known, has been previously outlined. The second involves modeling input processes by solving differential equations using filter techniques, where the input is a white noise process—referred to as the *pre-filter* technique.

The first method is advantageous when the input closely aligns with a shot noise process, providing accurate representation. However, this representation is limited as many real-world phenomena exhibit noticeable frequency modulation, making it suitable only in specific cases.

The second approach is more versatile and capable of representing phenomena with varying frequency contents, even those changing over time. This flexibility is crucial for accurately describing phenomena that could lead to resonant effects in structures. Additionally, this method retains the advantages associated with shot noise inputs, making it the preferred choice for many real structural issues.

In particular, using the *pre-filter* approach, the filter response is described by the $2m_f$ filter space state vector, the solution of the $2m_f$ set of differential equations

$$\dot{\overline{Z}}_f = \mathbf{A}_f(t)\overline{Z}_f + \mathbf{G}_f\overline{W}_f(t) \tag{9}$$

that generally could have a time-dependent form, when not only the frequency but also the amplitude of loads has an intrinsic evolutive nature. $\overline{W}$ is a vector of n_f white noise processes (stationary or non-stationary), $\mathbf{G}_f$ is a $m_{fx}n_f$ matrix that couples the excitation components of the forcing vector to the filter degree of freedom, and finally, $\mathbf{A}_f(t)$ is the $2m_f$ X. $2m_f$ filter system matrix, whose generic form is

$$\mathbf{A}_f(t) = \begin{pmatrix} \mathbf{0} & \mathbf{I} \\ \mathbf{H}_f^1(t) & \mathbf{H}_f^2(t) \end{pmatrix} \tag{10}$$

Then, adopting the *pre-filter* technique, the motion differential equations are written in the space state as

$$\dot{\overline{Z}}_s(t) = \begin{pmatrix} \mathbf{0} & \mathbf{I} \\ -\mathbf{M}^{-1}\mathbf{K} & -\mathbf{M}^{-1}\mathbf{C} \end{pmatrix} \overline{Z_s}(t) + \boldsymbol{\alpha}(t)\overline{Z_f}(t) \tag{11}$$

$$\dot{\overline{Z}}_f = \mathbf{A}_f(t)\overline{Z}_f + \mathbf{G}_f\overline{W_f}(t) \tag{12}$$

where

$$\mathbf{A}_s = \begin{pmatrix} \mathbf{0} & \mathbf{I} \\ \mathbf{H}_s^1 & \mathbf{H}_s^2 \end{pmatrix} \tag{13}$$

is the structural system matrix

$$\boldsymbol{\alpha}(t) = \begin{pmatrix} \mathbf{0} & \mathbf{0} \\ \mathbf{G}_s\boldsymbol{\alpha}_1(t) & \mathbf{G}_s\boldsymbol{\alpha}_2(t) \end{pmatrix} \tag{14}$$

is a $2n_s x 2m_s$ time-dependent matrix

$$\overline{Z}_s = \begin{pmatrix} \overline{X}_s & \dot{\overline{X}}_s \end{pmatrix}^T \tag{15}$$

is the structural space vector. Equations for the space state structure can be summarized as

$$\dot{\overline{Z}}(t) = A(t)\overline{Z}(t) + G(t)\overline{W_f}(t) \tag{16}$$

where

$$\overline{Z}^T = \left(\overline{X}_s, \overline{X}_f, \dot{\overline{X}}_S, \dot{\overline{X}}_f\right)^T \tag{17}$$

$$\mathbf{G}(t) = \begin{pmatrix} \mathbf{0} \\ \mathbf{G}_f \end{pmatrix} \tag{18}$$

is a new global $2m = 2\left(m_s + m_f\right)$ space state vector (structure plus filter) and

$$\frac{d}{dt}\begin{pmatrix} \overline{X}_s \\ \overline{X}_f \\ \dot{\overline{X}}_S \\ \dot{\overline{X}}_f \end{pmatrix} = \begin{pmatrix} 0 & 0 & I & 0 \\ 0 & 0 & 0 & I \\ H_s^1 & G_s\alpha_1(t) & H_s^2 & G_s\alpha_2(t) \\ 0 & H_f^1(t) & 0 & H_f^1(t) \end{pmatrix}\begin{pmatrix} \overline{X}_s \\ \overline{X}_f \\ \dot{\overline{X}}_S \\ \dot{\overline{X}}_f \end{pmatrix} + \begin{pmatrix} \overline{0} \\ \overline{0} \\ \overline{0} \\ G_f\overline{W}_f \end{pmatrix} \tag{19}$$

The structural matrix and response vectors are contingent on the design parameter vector $\overline{b}$, which encompasses elements such as structural stiffness, damping, masses, and various mechanical parameters like cross-sections, Young's modulus, and boundary conditions, among others. Filter parameters and input intensity are also included within this set of design parameters. Consequently, the system matrix and equations, comprising both the space state structure and filter equation, can be explicitly reconfigured as a function of this design parameter vector:

$$\mathbf{A}(b,\,t) = \begin{pmatrix} \mathbf{0} & \mathbf{0} & \mathbf{I} & \mathbf{0} \\ \mathbf{0} & \mathbf{0} & \mathbf{0} & \mathbf{I} \\ \mathbf{H}_s^1(\overline{b}) & \mathbf{G}_s\boldsymbol{\alpha}_1(\overline{b},\,t) & \mathbf{H}_s^2(\overline{b}) & \mathbf{G}_s\boldsymbol{\alpha}_2(\overline{b},\,t) \\ \mathbf{0} & \mathbf{H}_f^1(\overline{b},\,t) & \mathbf{0} & \mathbf{H}_f^1(\overline{b},\,t) \end{pmatrix} \tag{20}$$

$$\dot{\overline{Z}}(b,\,t) = A(\overline{b},\,t)\overline{Z}(\overline{b},\,t) + G(\overline{b},\,t)\overline{W_f}(t) \tag{21}$$

3. Space State Covariance Evaluation

In case of zero initial conditions, the solution of (space state structure + filter equation) has the following general expression:

$$\overline{Z}(t) = \int_0^t \boldsymbol{\Phi}(t,\,\tau)G\overline{W_f}(t)d\tau \tag{22}$$

where the matrix $\boldsymbol{\Phi}(t_1,\,t_2)$ (see for example [3]) is usually called *transition matrix*. The mean space state vector $\overline{\mu}_z(t)$ could be determined by the differential vectorial equation:

$$\dot{\overline{\mu}}_z\left(\overline{b},\,t\right) = A\overline{\mu}_z\left(\overline{b},\,t\right) + G\overline{\mu}_w(t) \tag{23}$$

The covariance matrix is as follows:

$$\mathbf{R}_{ZZ}(t_1,\,t_2) = \begin{pmatrix} \mathbf{R}_{XX}(t_1,\,t_2) & \mathbf{R}_{X\dot{X}}(t_1,\,t_2) \\ \mathbf{R}_{\dot{X}X}(t_1,\,t_2) & \mathbf{R}_{\dot{X}\dot{X}}(t_1,\,t_2) \end{pmatrix} \tag{24}$$

This second-order statistical moments matrix, due to its symmetry, is described by $(2m^2 + m)$ independent elements and can be evaluated by the well-known Lyapunov differential matrix equation

$$\dot{R}_{ZZ}(\overline{b},\,t) = A(\overline{b},\,t)R_{ZZ}(\overline{b},\,t) + R_{ZZ}(\overline{b},\,t)A(\overline{b},\,t)^T + B(t) \tag{25}$$

where

$$\mathbf{B}(t) = \left\langle \overline{Z}(t)\mathbf{G}^T\overline{W}(t)^T \right\rangle + \left\langle \mathbf{G}\overline{W}(t)\overline{Z}^T(t) \right\rangle = \mathbf{P}(t) + \mathbf{P}^T(t) \tag{26}$$

and where $\mathbf{P}$ can be written as

$$\mathbf{P}(t) = \int_0^t \boldsymbol{\Phi}(t-\tau)\mathbf{G}\left\langle \overline{W}(\tau)\overline{W}^T(t) \right\rangle \mathbf{G}^T d\tau = \int_0^t \boldsymbol{\Phi}(t-\tau)\mathbf{N}(t,\,\tau)d\tau \tag{27}$$

and

$$\mathbf{N}(t,\,\tau) = \mathbf{G}\mathbf{R}_{ww}(t,\,\tau)\mathbf{G}^T. \tag{28}$$

It must be noticed that Equation (25), which is valid for the non-stationary case, in a stationary environmental situation, is simplified in the following equation that contains no more time dependency:

$$A\left(\overline{b}\right)R_{ZZ}\left(\overline{b}\right) + R_{ZZ}\left(\overline{b}\right)A\left(\overline{b}\right)^T + B = 0 \tag{29}$$

A serious simplification takes place when the forcing vector is a white noise process as defined in (Rww stationary) or (Rww non-stationary). In these cases, the matrix $\mathbf{B}(t)$ is equal to

$$\mathbf{B}(t) = \left(\int_0^t \mathbf{\Phi}(t-\tau)\mathbf{G}\mathbf{R}_{ww}(t,\tau)\mathbf{G}^T d\tau\right) + \left(\int_0^t \mathbf{\Phi}(t-\tau)\mathbf{G}\mathbf{R}_{ww}(t,\tau)\mathbf{G}^T d\tau\right)^T \quad (30)$$

where both integrals above are equal due to the Dirac function properties. Finally, $\mathbf{B}$ can be written as follows:

$$\mathbf{B}(t) = 2\mathbf{G}\mathbf{R}_{ww}(t,t)\mathbf{G}^T = \begin{bmatrix} \mathbf{0}^{mxm} & \mathbf{0}^{mxm} \\ \mathbf{0}^{mxm} & \mathbf{L}(t) \end{bmatrix} \quad (31)$$

where the MXM submatrix is diagonal with the elements

$$[\mathbf{L}(t)]_{k,k} = \frac{2\pi S_{0_k}}{m_k^2}\varphi_k^2(t) \quad (32)$$

Meanwhile, for some applications, covariance information about structural acceleration is needed so that the matrix

$$\mathbf{R}_{\overline{\ddot{X}\ddot{X}}}(b,t) = \left\langle \overline{\ddot{X}\ddot{X}}^T \right\rangle \quad (33)$$

must be determined, and it is easily obtainable by the relation

$$R_{\overline{\ddot{X}\ddot{X}}}(\bar{b},t) = D(b,t)R_{\overline{ZZ}}(b,t)D(b,t)^T \quad (34)$$

where

$$D(b,t) = \begin{bmatrix} \mathbf{H}_s^1(b) & \mathbf{G}_s\boldsymbol{\alpha}_1(b,t) & \mathbf{H}_s^2(b) & \mathbf{G}_s\boldsymbol{\alpha}_2(b,t) \end{bmatrix}. \quad (35)$$

4. Time Integration BF Procedure for R and $\mathbf{R}_{,\bar{b}}$

Even if many numerical standard codes exist for the stationary Lyapunov equation (the Lyapunov equation in stationary conditions is where A and B are the input matrices and X is the unknown one), there are limited examples for addressing the non-stationary Lyapunov equation, and so a simple numeric implicit integration method is proposed. In this context, a straightforward numerical implicit integration method is suggested. Specifically, the modified *Euler* method is used, in which the period is divided in m equals time steps d in each sub-period Δt, and a linear variation in the time derivative covariance matrix $\dot{R}(t)$ is assumed. Under this assumption, we have the (*standard implicit Euler method*):

$$R^{(h+1)} = R^{(h)} + \frac{1}{2}\Delta t\left[\dot{R}^{(h+1)} + \dot{R}^{(h)}\right] \quad (36)$$

where the symbol $a^{(h)}$ denotes the generic quantity a evaluated at time $t = h\Delta t$. By using the matrix equations evaluated at times $t^{(h+1)}$ and $t^{(h)}$, we obtain the following m algebraic matrix equations of the Lyapunov type

$$\begin{gathered}\left[\left(\tfrac{1}{2}(\mathbf{I}-\Delta t\mathbf{A})\right)\mathbf{R}^{(h+1)} + \mathbf{R}^{(h+1)}\left(\tfrac{1}{2}(\mathbf{I}-\Delta t\mathbf{A})\right)^T\right] = \\ \left[\left(\tfrac{1}{2}(\mathbf{I}+\Delta t\mathbf{A})\right)\mathbf{R}^{(h)} + \mathbf{R}^{(h)}\left(\tfrac{1}{2}(\mathbf{I}+\Delta t\mathbf{A})\right)^T\right] + \tfrac{\Delta t}{2}\left(\mathbf{B}^{(h)} + \mathbf{B}^{(h+1)}\right)\end{gathered} \quad (37)$$

that are solved in sequence for each time value, starting from the initial time value and the initial covariance matrix value.

In this way, the m unknown matrices $(h = 1, \ldots, m)$ are determined. By assuming a constant or time variable (depending on the filter parameters variation), the matrices are

$$\begin{aligned}\mathbf{P}_B &= \tfrac{1}{2}(\mathbf{I}-\Delta t\mathbf{A}) \\ \mathbf{P}_F &= \tfrac{1}{2}(\mathbf{I}+\Delta t\mathbf{A})\end{aligned} \quad (38)$$

and we discern that a more compact form of (Rnum1) is

$$\left[\mathbf{P}_B\mathbf{R}^{(h+1)} + \mathbf{R}^{(h+1)}\mathbf{P}_B^T\right] = \left[\mathbf{P}_F\mathbf{R}^{(h)} + \mathbf{R}^{(h)}\mathbf{P}_F^T\right] + \frac{\Delta t}{2}\left(\mathbf{B}^{(h)} + \mathbf{B}^{(h+1)}\right) \tag{39}$$

that could be solved at each step via a standard stationary Lyapunov equation solver, for example, the leap in the standard Matlab toolbox, in the form

$$\mathbf{P}_B\mathbf{R}^{(h+1)} + \mathbf{R}^{(h+1)}\mathbf{P}_B^T + \mathbf{C}^{(h+1)} = \mathbf{0} \tag{40}$$

where $\mathbf{C}^{(h+1)} = -\left(\left[\mathbf{P}_F\mathbf{R}^{(h)} + \mathbf{R}^{(h)}\mathbf{P}_F^T\right] + \frac{\Delta t}{2}\left(\mathbf{B}^{(h)} + \mathbf{B}^{(h+1)}\right)\right)$.

Covariance could be integrated as proposed in the following integration scheme (Algorithm 1), where n_t is the number of time integration steps, T is the total analysis time, n_b the number of design parameters, and, for the sake of simplicity in notation, $\mathbf{S}(h,\, j) = \mathbf{R}_{,b_j}((h-1)\,\Delta t)$:

Algorithm 1: Integration scheme

% *data input* $\left(\mathbf{M}, \mathbf{C}, \mathbf{K}, S_0, \omega_f, \xi_f, \phi(t)\right)$ %

$\mathbf{P}_B = \frac{1}{2}(\mathbf{I} - \Delta t\mathbf{A})$

$\mathbf{P}_F = \frac{1}{2}(\mathbf{I} + \Delta t\mathbf{A})$

$\mathbf{R}(1) = \mathbf{0}$; *(definition of initial condition for t=0 both on* $\mathbf{R}$ *and on* $\mathbf{R}_{,b}$*)*

for $j=1$ *to* n_b % *initial condition definition*%

$\mathbf{S}(j,1) = 0$;

end;

for $i=1$ *to* $(n_t - 1)$ % *beginning of time integration* %

$\mathbf{N}(i) = \frac{\Delta t}{2}(\mathbf{B}(i+1) + \mathbf{B}(i))$

$\mathbf{C}(i) = \mathbf{P}_F\mathbf{R}(i) + \mathbf{R}(i)\mathbf{P}_F^T + \mathbf{N}(i)$

$\mathbf{R}(i+1) =$ leap $(\mathbf{P}_B, -\mathbf{C}(i))$

standard solution for stationary Lyapunov equation $\mathbf{R}(i+1)\mathbf{P}_B + \mathbf{P}_B\mathbf{R}^T(i+1) - \mathbf{C}(i) = 0$.

for $j=1$ *to* n_b *cycle for each design vector element*

$\mathbf{M}(i,j) = \frac{\Delta t}{2}\left[\mathbf{A}_b(j)(\mathbf{R}(i) + \mathbf{R}(i+1)) + (\mathbf{R}(i) + \mathbf{R}(i+1))\mathbf{A}_b^T(j)\right]$

$\mathbf{D}(i,j) = \mathbf{P}_F\mathbf{S}(i,j) + \mathbf{S}(i,j)\mathbf{P}_F^T + \mathbf{M}(i,j)$

$\mathbf{S}(i+1,j) =$ lyap $(\mathbf{P}_B, \mathbf{D}(i,j))$

end % *end of cycle on design vector elements* %

end % *end of time integration* %

5. Numerical Example

The method proposed is used to analyze a multi-story building under earthquake forces. To maintain a balance between simplicity and generality, a shear-type plane frame structure is chosen as the model (as shown in Figure 1). This choice is reasonable because, in many buildings, the floor slabs possess very high in-plane stiffness, allowing them to be treated as rigid diaphragms. This simplification significantly enhances analysis efficiency without substantially compromising the accuracy of response assessment to ground forces.

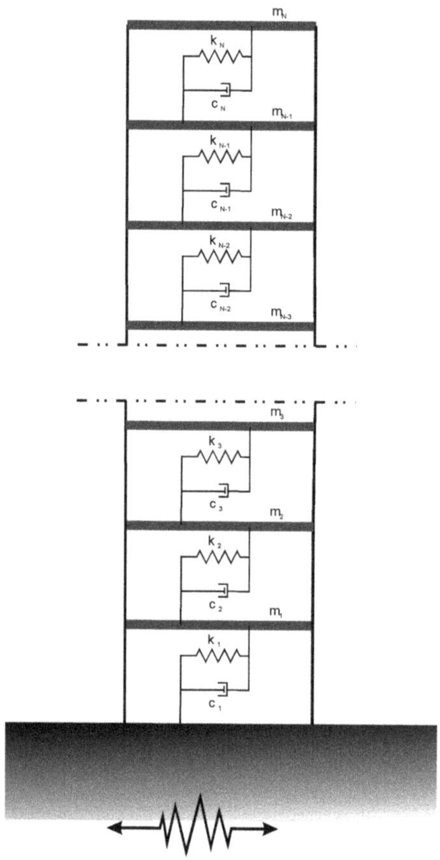

Figure 1. Mechanical scheme of analyzed plane frame shear type.

Moreover, to further enhance computational efficiency, the matrix condensation technique is employed. The primary assumption in modeling the building's mechanics is linearity, which remains valid considering the limitations on horizontal displacements necessary for operational service levels. This assumption holds when the maximum inter-story drift approaches or reaches the elastic limit of structural displacements, typical in full operational design demands.

For the sensitivity analysis, the design vector comprises the masses, stiffness, and damping of each floor. This comprehensive vector allows for the evaluation of how variations in these parameters impact the final structural reliability.

$$\bar{b} = (\bar{b}_1, \bar{b}_2, \bar{b}_3) = \left(\overline{m}^T, \bar{k}^T, \bar{c}^T\right)$$

where $\bar{b}_2 = \bar{k}^T = (k_1, k_2, k_3, \ldots, k_n)$.

5.1. Equations of Motion

To capture the essential seismic characteristics involving spectral and time modulation, a non-stationary modulated Kanai–Tajimi process is utilized to model stochastic ground motion. This process characterizes the base acceleration $\ddot{X}_g(t)$ acting at the structure's base as follows:

$$\begin{cases} \ddot{X}_g(t) = \ddot{X}_f(t) + \phi(t)w(t) \\ \ddot{X}_f(t) + 2\xi_g\omega_g\dot{X}_f(t) + \omega_g^2 X_f(t) = -\phi(t)w(t) \end{cases}$$

where $X_f(t)$ is the response of the Kanai–Tajimi filter, with a frequency ω_g and damping coefficient ξ_g, and $w(t)$ is the white noise, whose constant bilateral power spectral density (PDS) function is S_0. This last parameter is related to the peak ground acceleration (PGA) $\ddot{X}_g^{\max}$ by means of relation [16]

$$S_0 = 0.2222\frac{\xi_g\left(\ddot{X}_g^{\max}\right)^2}{\pi\omega_g\left(1+4\xi_g^2\right)}$$

The non-stationary nature is introduced through the deterministic temporal modulation function $\phi(t)$, regulating the intensity variations while preserving the earthquake's frequency characteristics. In this scenario, the specific modulation functions proposed by Jennings [17] are adopted:

$$\varphi(t) = \begin{cases} \left(\frac{t}{t_1}\right)^2 & t < t_1 \\ 1 & t_1 \le t \le t_2 \\ e^{-\beta(t-t_2)} & t > t_2 \end{cases} \tag{41}$$

The motion equations for the complete structural system are then as follows:

$$\begin{gathered} \ddot{\overline{X}}(t) + M(^{\overline{b}}C(\overline{b})\dot{\overline{X}}(t) + M(^{\overline{b}}K(\overline{b})\overline{X}(t) = \overline{r}\ddot{X}_g(t) \\ \ddot{X}_f(t) + 2\xi_f\omega_f\dot{X}_f(t) + \omega_f^2 X_f(t) = -w(t)\phi(t) \end{gathered}$$

where the drag vector $nx1$ is $\overline{r} = [1, 1, 1, \ldots, 1, 1]^T$, $\mathbf{M}(\overline{b})$, $\mathbf{C}(\overline{b})$, and $\mathbf{K}(\overline{b})$ are, respectively, mass, viscosity, and stiffness $n_x n$ principal structure matrices, whose general expression are reported in Appendix A with reference to the rigid floor assumption. The three vectors $\ddot{\overline{X}}(\overline{b}, t)$, $\dot{\overline{X}}(\overline{b}, t)$, and $\overline{X}(\overline{b}, t)$ are ground relative acceleration, velocity, and displacement $n_x 1$ vectors. In the case of the analyzed structure, the mass matrix is diagonal and the two viscous and stiffens matrices are tri-diagonal once. The mechanical filter parameters, the damping ratio and frequency, are ξ_f and ω_f, and the base excitation $\ddot{X}_g(t)$ is then equal to $\phi(t)w(t) + \ddot{X}_f(t)$.

Introducing the state vector $\overline{Y}(\overline{b}, t) = \left\{\overline{X}^T(\overline{b}, t), X_f(t)\right\}^T$, the motion equation system (motion Equation (1)) can be rewritten as

$$\ddot{\overline{Y}}(\overline{b}, t) = -H_1(\overline{b})\dot{\overline{Y}}(\overline{b}, t) - H_2(\overline{b})\overline{Y}(\overline{b}, t) - \overline{f}(t)$$

where the two matrices $\mathbf{H}_1(\overline{b})$ and $\mathbf{H}_2(\overline{b})$, with the vector $\overline{f}(t)$, are defined for this specific problem and shown in Appendix B.

The $N = n + 1$ degree of freedom 2nd-order differential system (second-order moment equals complete) can be replaced with a $2N$ DoF 1st-order differential equation in the space state as follows:

$$\dot{\overline{Z}}(\overline{b}, t) = A(\overline{b})\overline{Z}(\overline{b}, t) + \overline{F}(t)$$

where the space vector $2N$ is $\overline{Z}(\overline{b}, t) = \left\{\overline{Y}(\overline{b}, t), \dot{\overline{Y}}(\overline{b}, t)\right\}$ and the *system matrix* ($2N$ X. $2N$) $\mathbf{A}(\overline{b}) = \begin{pmatrix} \mathbf{0}^{n+1} & \mathbf{I}^{n+1} \\ -\mathbf{H}_2(\overline{b}) & -\mathbf{H}_1(\overline{b}) \end{pmatrix}$ and the $2N$ forcing vector $\overline{F}(t) = \begin{Bmatrix} \overline{0}^{n+1} \\ \overline{f}(t) \end{Bmatrix}$.

5.2. Reliability Evaluation

With reference to the proposed problem, it is required to evaluate the probability that each story drift U_h of the floor exceeds the thresholds, at least once in a given earthquake duration. Then, for each hth level, this failure event is associated with the condition $|x_{h+1} - x_h| = |u_h| = \beta_h$. For each level h, the reliability vector element $r_h(\overline{b}, T)$ is defined as where, under the Poisson hypothesis for threshold crossing (that is an acceptable hypothesis for rare events as in [18]), we obtain

$$r_h(\overline{b}, \beta_h, T) =$$
$$\exp\left\{-\frac{1}{\pi}\int_0^T \left(\frac{\sigma_{\dot{U}_h}(\overline{b},\tau)}{\sigma_{U_h}(\overline{b},\tau)}\sqrt{1-\rho^2_{U_h\dot{U}_h}(\overline{b}, t)}\exp\left\{-\frac{1}{2}\eta_h^2(\overline{b}, \beta_h, \tau)\right\}\chi\left[d_{U_h}(\overline{b}, \beta_h, t)\right]\right)d\tau\right\}$$

and the final global structural reliability is then

$$r_{global}(\overline{b}, \overline{\beta}, t) = \prod_{h=1}^{n} r_h\left(\overline{b}, \beta_h, t\right)$$

Although exact analytical solutions for this are generally unavailable, it is known that the equation (approximate upper-bound global reliability) provides an approximate, upper-bound estimate, as stated above, and can be used for design and pre-design purposes from a practical viewpoint, as in this study. In order to evaluate the reliability vector (adopting the Poisson approach) related to the inter-floor relative displacement threshold crossing, one needs to introduce the inter-story drift vector with the associated covariance matrix $R_{Z_U Z_U}(t)$ (see Appendix C).

The reliability vector $\overline{r}_U$ previously defined can be evaluated as the collection of

$$r_{Uh}(T) = r_0\, e^{-\int_0^T v^+_{\dot{U}h}(\tau)d\tau}$$

where $v^+_{\dot{U}_h}(\eta_{V_h})$ is a function of $\sigma^2_{\dot{U}_h}$, $\sigma^2_{\dot{U}_h}$, and $\rho_{U_h\dot{U}_h}$, accordingly, and are, respectively, the h and the $n+h$ diagonal elements of $\mathbf{R}_{Z_U Z_U}(t)$ and $\eta_{U_h} = \frac{\beta_h}{\sigma_{U_h}}$, with β_h being the hth barrier.

The equation represents the probability that the inter-story displacement will cross the maximum acceptable value β_h during the time interval $[0, T]$. All the barriers can be collected in the barrier vector. Its elements are assumed constant and equal to 3.0 cm for each floor, that is, there is a lateral drift equal to 1.0% in the case of inter-story height of 3 m.

5.3. System Parameters

The chosen building configuration consists of three stories, each with a uniform mass equal to x 10^5 (kg) for each level. Lateral stiffness to the first floor is present, and it is assumed that a linear reduction decreases its value to the at the top floor ($k_2 = 5.1 \cdot 10^7$(N/m) and $k_3 = 4.2 \cdot 10^7$(N/m)). Finally, damping is evaluated by setting $c_i = 2\sqrt{m_i k_i}$ (c_i = [3.0 2.8 2.5] 10^5(N s/m)).

The seismic characterization is based on a peak ground acceleration (PGA) of 0.45 (g), and four distinct total durations (10, 20, 30, and 40 s) are employed. Figure 2 displays the structural inter-story covariances, measured in terms of displacement (a) and velocity (b). It is noteworthy that for durations exceeding this, structural responses attain a stationary level. Moreover, for this particular structural configuration, the covariance response of the first level is greater than the other two, even though the third is approximately half, while the second is only slightly smaller.

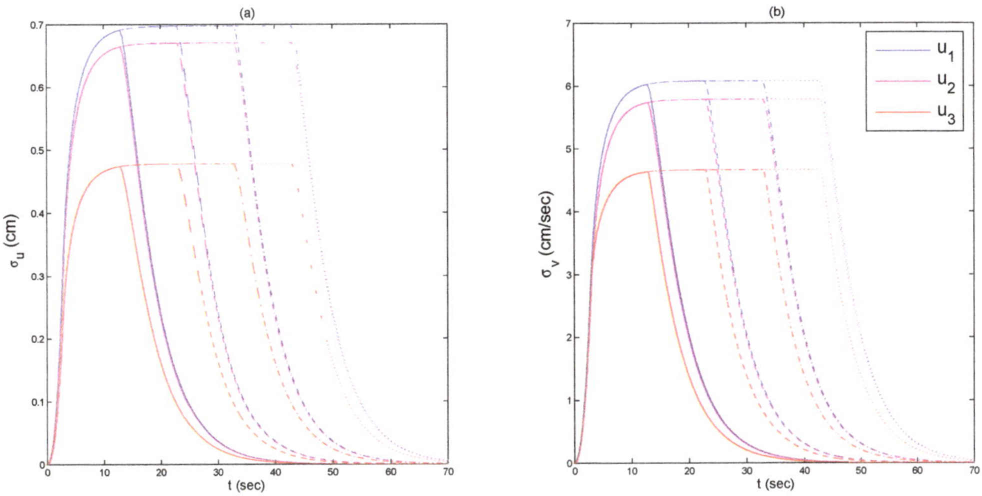

Figure 2. Inter-story displacements (**a**) and velocities (**b**) covariance response of a 3DoF system subject to modulated filtered white noise with different time durations. Continuous lines are for t_{d} = 10 (s), dashed lines are for t_{d} = 20 (s), dash–dot lines are for t_{d} = 30 (s), and finally, dotted lines are for t_{d} = 40 (s). Blue lines represent the inter-story drift response on the first floor, magenta lines on the second floor, and red lines refer to the third floor.

Figure 3 illustrates the structural safeties assessed at each lateral inter-story drift threshold for failure, along with the overall reliability calculated as an approximate upper-bound global reliability. It is important to observe that the probability of failure is highest for the first inter-story drift threshold, followed by a slightly lower probability for the second one, and finally, the third threshold has a considerably negligible probability of failure (i.e., $r_3 = 1$).

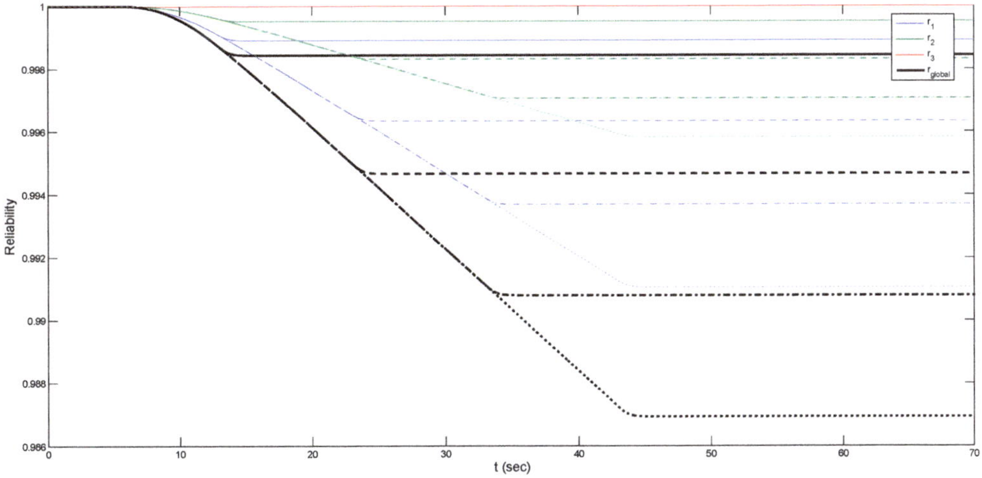

Figure 3. System reliability of a 3DoF system, evaluated as the probability of maximum inter-story drift exceeds a given threshold of 3 (cm). Results are obtained for different values of t_{d}: continuous lines are for t_{d} = 10 (s), dashed lines are for t_{d} = 20 (s), dash–dot lines are for t_{d} = 30 (s), and finally, dotted lines are for t_{d} = 40 (s). Blue lines are for first-level reliability, magenta lines are for second-level reliability, and red lines are for third-level reliability. Black slight lines are for global system reliability.

6. Conclusions

A numerical time integration algorithm is proposed to deal with non-stationary random vibration problems utilizing a covariance approach. The algorithm is developed to solve the differential matrix equations that govern the evolution of the stochastic response of structures subjected to random inputs. To create a versatile non-stationary approach applicable in various contexts, the structural response is assessed through a covariance approach. The reliability concerning first-crossing failure events is then derived based on the knowledge of the evolving covariance matrix in the space state. The algorithm is proposed for a generic Gaussian input of filtered non-stationary processes, representing diverse real-world physical loads. To solve this problem, the time integration algorithm involves differentiating the Lyapunov equation using an adapted Euler implicit scheme, which can be easily implemented using standard tools in various programming codes. Finally, the proposed algorithm is applied to analyze the dynamic responses of a multistory building, idealized as a shear frame structure.

Author Contributions: Conceptualization, R.C. and G.C.M.; Methodology, M.D., R.C. and G.C.M.; Validation, R.C. and G.C.M.; Formal analysis, M.D., R.C. and G.C.M.; Investigation, G.C.M.; Writing—original draft, M.D., R.C. and G.C.M.; Writing—review & editing, M.D., R.C., S.L.L., M.M. and G.C.M.; Visualization, L.S.; Supervision, G.C.M. All authors have read and agreed to the published version of the manuscript.

Funding: The research leading to these results has received funding from the European Research Council under the grant agreement ID: 101007595 of the project ADDOPTML, MSCA RISE 2020 Marie Skodowska Curie Research and Innovation Staff Exchange (RISE).

Conflicts of Interest: The authors declare no conflict of interest.

Appendix A

Matrices, **C** and **K** are, for a shear-type frame, respectively diagonal and tri-diagonals:

$$\mathbf{M} = \begin{pmatrix} m_1 & 0 & 0 & & & & \\ 0 & m_2 & 0 & & & & \\ & 0 & m_3 & \ddots & & & \\ & & \ddots & \ddots & & & \\ & & & 0 & m_{n-1} & 0 \\ & & & & 0 & m_n \end{pmatrix} \tag{A1}$$

$$\mathbf{K} = \begin{pmatrix} k_1+k_2 & -k_2 & 0 & & & \\ -k_2 & k_2+k_3 & -k_3 & 0 & & \\ 0 & -k_3 & k_3+k_4 & \ddots & & \\ & 0 & \ddots & \ddots & -k_{n-1} & 0 \\ & & & -k_{n-1} & k_{n-1}+k_n & -k_n \\ & & & 0 & -k_n & k_n \end{pmatrix} \tag{A2}$$

$$\mathbf{C} = \begin{pmatrix} c_1+c_2 & -c_2 & 0 & & & \\ -c_2 & c_2+c_3 & -c_3 & 0 & & \\ 0 & -c_3 & c_3+c_4 & -c_4 & & \\ & 0 & \ddots & \ddots & \ddots & 0 \\ & & & -c_{n-1} & c_{n-1}+c_n & -c_n \\ & & & 0 & -c_n & c_n \end{pmatrix} \tag{A3}$$

Appendix B

Matrices $\mathbf{H}_1$ and $\mathbf{H}_2$ are

$$\mathbf{H}_1 = \begin{pmatrix} & & & & & & 2\xi_f\omega_f \\ & & & & & & 2\xi_f\omega_f \\ & & & & & & . \\ & & & \mathbf{M}^{-1}\mathbf{C} & & & . \\ & & & & & & 2\xi_f\omega_f \\ & & & & & & 2\xi_f\omega_f \\ 0 & 0 & 0 & \ldots\ldots & 0 & 0 & -2\xi_f\omega_f \end{pmatrix} \quad \text{(A4)}$$

$$\mathbf{H}_2 = \begin{pmatrix} & & & & & & \omega_f^2 \\ & & & & & & \omega_f^2 \\ & & & & & & . \\ & & & \mathbf{M}^{-1}\mathbf{K} & & & . \\ & & & & & & \omega_f^2 \\ & & & & & & \omega_f^2 \\ 0 & 0 & 0 & \ldots\ldots & 0 & 0 & -\omega_f^2 \end{pmatrix} \quad \text{(A5)}$$

where ω_f^2 and ξ_f are filter characteristics and the forcing vector is $\overline{f}(t) = [0, 0, 0, \ldots\ldots, 0, \varphi(t)w(t)]^T$.

Appendix C

The covariance matrix $R_{Z_U Z_U}\left(\overline{b}, t\right)$ is defined in the linear space of stochastic processes as a linear equation related to the inter-story drift $\overline{U}(\overline{b}, t)$ and displacement $\overline{X}(\overline{b}, t)$ vectors, holds $U(\overline{b}, t) = \mathbf{T}\overline{X}(\overline{b}, t)$ where the transform matrix $\mathbf{T}$ is a bi-diagonal one, and is independent of the design vector:

$$\mathbf{T} = \begin{pmatrix} 1 & 0 & 0 & 0 & & & & \cdots & 0 \\ -1 & 1 & 0 & 0 & & & & & \\ 0 & -1 & 1 & 0 & & & & & \vdots \\ & & -1 & 1 & \ddots & & & & \\ & & & & \ddots & 1 & 0 & 0 & \\ \vdots & & & & 0 & -1 & 1 & 0 & 0 \\ & & & & & 0 & -1 & 1 & 0 \\ 0 & & & & & \cdots & & 0 & 1 \end{pmatrix} \quad \text{(A6)}$$

The covariance matrix $\mathbf{R}_{Z_U Z_U}\left(\overline{b}, t\right)$ is then related to $\mathbf{R}_{ZZ}\left(\overline{b}, t\right)$ through the following connection:

$$\widehat{T} = \begin{pmatrix} T & 0 \\ 0 & T \end{pmatrix} \quad \text{(A7)}$$

$$\overline{Z}_V(\overline{b}) = \widehat{T}\overline{Z}(\overline{b}) \quad \text{(A8)}$$

$$R_{Z_U Z_U}(\overline{b}, t) = \widehat{T}R_{ZZ}(\overline{b}, t)\widehat{T}^T \quad \text{(A9)}$$

References

1. Liu, J.; Wang, L. Hybrid Reliability-Based Sequential Optimization for PID Vibratory Controller Design Considering Interval and Fuzzy Mixed Uncertainties. *Appl. Math. Model.* **2023**, *122*, 796–823. [CrossRef]
2. Huang, H.; Li, Y.; Li, W. Dynamic reliability analysis of stochastic structures under non-stationary random excitations based on an explicit time-domain method. *Struct. Saf.* **2023**, *101*, 102313. [CrossRef]

3. Soong, T.T.; Grigoriu, M. *Random Vibration of Mechanical and Structural Systems*; Prentice Hall: Hoboken, NJ, USA, 1993.
4. Lin, Y.K. *Probabilistic Theory of Structural Dynamics*; Wax, N., Ed.; Stochastic Processes; Krieger Publishing, Co., Inc.: Huntington, NY, USA, 1967.
5. Lutes, L.D.; Sarkani, S. *Stochastic Analysis of Structural and Mechanical Vibration*; Prentice-Hall Inc.: Hoboken, NJ, USA, 1997.
6. Nigam, N.C.; Narayanan, S. *Application of Random Vibration*; Narosa Publishing House: Delhi, India, 1994.
7. Gomez, F.; Spencer, B.F., Jr.; Carrion, J. Topology optimization of buildings subjected to stochastic base excitation. *Eng. Struct.* **2020**, *23*, 111111. [CrossRef]
8. Spencer, B.F.; Gomez, F.; Xu, J. *Topology Optimization for Stochastically Excited Structures*; ISSRI: Uttar Pradesh, India, 2016.
9. Kressner, D. Memory-efficient Krylov subspace techniques for solving large-scale Lyapunov equations. In Proceedings of the 2008 IEEE International Conference on Computer-Aided Control Systems, San Antonio, TX, USA, 3–5 September 2008; IEEE: Piscataway, NJ, USA, 2008; pp. 613–618.
10. Saad, Y. *Numerical Solution of Large Lyapunov Equations*; Signal Process Scattering Oper Theory Numer Methods, Proc MTNS-89; Birkhauser: Basel, Switzerland, 1989; Volume 3, pp. 503–511.
11. Jbilou, K.; Riquet, A. Projection methods for large Lyapunov matrix equations. *Linear Algebra Appl.* **2006**, *415*, 344–358. [CrossRef]
12. Muscolino, G.; Genovese, F.; Biondi, G.; Cascone, E. Generation of fully non-stationary random processes consistent with target seismic accelerograms. *Soil Dyn. Earthq. Eng.* **2021**, *141*, 106467. [CrossRef]
13. Marano, G.C.; Acciani, G.; Fiore, A.; Abrescia, A. Integration algorithm for covariance Nonstationary dynamic analysis of SDOF systems using equivalent stochastic linearization. *Int. J. Struct. Stab. Dyn.* **2015**, *15*, 1450044. [CrossRef]
14. Greco, R.; Marano, G.C.; Fiore, A.; Vanzi, I. Nonstationary First Threshold Crossing Reliability for Linear System Excited by Modulated Gaussian Process. *Shock. Vib.* **2018**, *2018*, 3685091. [CrossRef]
15. Sgobba, S.; Stafford, P.J.; Marano, G.C.; Guaragnella, C. An evolutionary stochastic ground-motion model defined by a seismological scenario and local site conditions. *Soil Dyn. Earthq. Eng.* **2011**, *31*, 1465–1479. [CrossRef]
16. Crandall, S.H.; Mark, W.D. *Random Vibration in Mechanical Systems*; Academic Press: Cambridge, MA, USA, 1963.
17. Jennings, P.C. Periodic response of a general yielding structure. *J. Eng. Mech. Div.* **1964**, *90*, 131–166. [CrossRef]
18. Vanmarcke, E.H. On the Distribution of the First-Passage Time for Normal Stationary Random Processes. *ASME J. Appl. Mech. March* **1975**, *42*, 215–220. [CrossRef]

computation

MDPI

Article

Taylor Polynomials in a High Arithmetic Precision as Universal Approximators

Nikolaos Bakas [1,2]

[1] Computation-Based Science and Technology Research Center, The Cyprus Institute, 20 Konstantinou Kavafi Str., 2121 Nicosia, Cyprus; n.bakas@cyi.ac.cy

[2] Department of Information Technology & AI Lab, American College of Greece, Deree. 6 Gravias Str., Aghia Paraskevi, 15342 Athens, Greece; nbakas@acg.edu

Abstract: Function approximation is a fundamental process in a variety of problems in computational mechanics, structural engineering, as well as other domains that require the precise approximation of a phenomenon with an analytic function. This work demonstrates a unified approach to these techniques, utilizing partial sums of the Taylor series in a high arithmetic precision. In particular, the proposed approach is capable of interpolation, extrapolation, numerical differentiation, numerical integration, solution of ordinary and partial differential equations, and system identification. The method employs Taylor polynomials and hundreds of digits in the computations to obtain precise results. Interestingly, some well-known problems are found to arise in the calculation accuracy and not methodological inefficiencies, as would be expected. In particular, the approximation errors are precisely predictable, the Runge phenomenon is eliminated, and the extrapolation extent may a priory be anticipated. The attained polynomials offer a precise representation of the unknown system as well as its radius of convergence, which provides a rigorous estimation of the prediction ability. The approximation errors are comprehensively analyzed for a variety of calculation digits and test problems and can be reproduced by the provided computer code.

Keywords: function approximation; approximation errors; interpolation; extrapolation; numerical differentiation; numerical integration; ordinary differential equation; partial differential equation; system identification; inverse problems; Taylor series; Taylor polynomials

1. Introduction

The utilization of a high arithmetic precision (HAP) for the modeling of an unknown function exhibited a remarkable extrapolation ability in [1], with extrapolation spans 1000% higher than the existing methods in the literature. The basis of this method was the approximation of an unknown analytic function with a high arithmetic precision. This is an essential problem in a variety of numerical methods. Standard programming languages are limited to 16–64 floating point digits, and researchers have been taking into account a high arithmetic precision for the various computations regarding the numerical integration [2], interpolation [3], and solution of Partial Differential Equations (PDEs) [4].

Recent research works highlight the importance of a HAP in computations. In [5], the Clarinet framework is proposed to replace floating point arithmetic in various linear algebra and computer vision calculations. The effect of round-off errors when utilizing a standard accuracy for reduction algorithms is highlighted in [6], and a high-precision "RingAllreduce" algorithm was proposed. A high-precision ray-tracing algorithm is presented in [7], reducing round-off errors in the numerical examples. A high arithmetic precision is also significant in the design of Field Programmable Gate Arrays (FPGAs), and a new representation to tackle programming challenges is proposed in [8]. The GNU Multiple Precision Arithmetic Library (GMP) [9] is a widely used library in many computer languages, like C++, Python, and Julia, and a framework to enable its usage by Java was recently developed [10].

Nevertheless, standard techniques exist for performing interpolation with Taylor polynomials [11,12], as well as the solution of differential equations [13–15]. However, certain problems occur when applying these methods for scientific computing tasks, such as the well-known Runge phenomenon [16,17], which remains a major complication [18–20]. Taylor series arise in the foundations of differential calculus [21] by associating the behavior of a function around a point x_0 with its derivatives at that particular point.

Accordingly, Taylor series are capable of approximating any analytic function in theory. However, in the practice of computing, they often fail, and researchers use other approximators than Taylor polynomials, such as radial basis functions, Lagrange polynomials, Chebyshev polynomials, artificial neural networks, etc., to avoid numerical instabilities. A variety of numerical methods have been developed for such operations, as researchers have been observing that Taylor polynomials do not offer stable calculations. Utilizing a high arithmetic precision, we demonstrate that such need, which arose to address the computational inaccuracies, does not exist. Taking into account the high extrapolation spans attained in [1] and obtained with integrated radial basis functions [22,23] and some hundreds or even thousands of digits for the calculations, we apply a high arithmetic precision utilizing the "BigFloat" structure of Julia language [24], using the GMP [9] library to truncate the Taylor series, known as Taylor polynomials or partial sums.

The purpose of this work is to present a unified approach to interpolation, extrapolation, numerical differentiation, solution of partial differential equations, system identification, and numerical integration for problems which comprise given data of an unknown analytic function or the source for PDEs. The paper is organized as follows: the formulation of our approach is presented in Section 2; some basic operations and results for 1-dimensional interpolation, extrapolation, numerical differentiation, numerical integration, and solutions of ordinary differential equations are presented in Section 3; the results of multidimensional function approximation, solution of partial differential equations, and system identification are presented in Section 4; and the conclusions follow in Section 5.

Taylor polynomials provide a fundamental means to approximate complex functions and understand their behavior, such as rate of change, curvature, and higher-order characteristics. However, when utilizing standard floating-point precision, a variety of numerical methods fail to produce robust results, and researchers have been developing complex numerical methods and techniques to tackle numerical instabilities. Interestingly, when utilizing hundreds of digits of precision, the accuracy obtained is exceptional in a variety of computational tasks while keeping a unified, fundamental, straightforward, and interpretable representation with Taylor polynomials.

2. Description of the Method

Let $f(x)$ be an analytic function, which is unknown. It is given that the function takes values $\mathbf{f} = \{f_1, f_2, \ldots, f_N\}$ at specified points $\mathbf{x} = \{x_1, x_2, \ldots, x_N\}$ as in Figure 1 for a generic analytic function. By applying the Taylor series [21] of the function at some point x_0, we may write $f(x \pm x_0) = f(x_0) \pm \frac{f'(x_0)}{1!}(x - x_0) + \frac{f''(x_0)}{2!}(x - x_0)^2 \pm \cdots \pm \frac{f^{(n)}(x_0)}{n!}(x - x_0)^n + \cdots$. The derivatives of the function, $\mathbf{df} = \left\{f^0, f', f'', \ldots, f^{(n)}\right\}$, at x_0, divided by the corresponding factorial $n!$, are constant quantities. Hence, by truncating the series at the nth power, we derive that $f(x \pm x_0) \cong f(x_0) \pm \frac{f'(x_0)}{1!}(x - x_0) + \frac{f''(x_0)}{2!}(x - x_0)^2 \pm \cdots + \frac{f^{(n)}(x_0)}{n!}(x - x_0)^n + R_n(x)$, while the remainder of the approximation is bounded by $|R_n(x)| \leq \frac{f^{n+1}(x)}{(n+1)!}|x - x_0|^{n+1}, \forall x : |x - x_0| \leq r$ [25].

For a series$f(x) = \sum_{n=0}^{\infty} a_n(x - x_0)^n$, we have that the radius of convergence [25] r is a non-negative real number or ∞ such that the series converges if $|x - x_0| < r$ and diverges if $|x - x_0| \geq r$. That is to say, the series converges in the interval $(x_0 - r, x_0 + r)$. We may compute r using the ratio test, $\limsup|a_{n+1}/a_n|$, or using the root test, with $r = 1/\limsup_{n\to\infty} \sqrt[n]{|a_n|}$. We select the root test because the coefficients a_i often contain zero elements, making the division computationally unstable. Furthermore, because $\liminf(a_{n+1}/a_n) \leq \liminf((a_n)^{(1/n)}) \leq \limsup((a_n)^{(1/n)}) \leq \limsup(a_{n+1}/a_n)$ [26], the computed r from the root test is more precise, as it is bounded by the ratio test.

A high arithmetic precision was found capable of achieving an accurate computation of r for a known series, whereas the floating point fails. This is a significant part of the proposed numerical schemes, as the identification of r offers information on the larger disk where the series converges. Accordingly, we obtain knowledge of the interpolation accuracy or even the extrapolation span of the approximated function beyond the given domain. In particular, at $x_0 = 0$, we may write that

$$f(x) \cong a_0 \pm a_1 x + a_2 x^2 \pm \cdots + a_n x^n \tag{1}$$

where $\mathbf{a} = \left\{1, f'/1, f''/2!, \ldots, f^{(n)}/n!\right\} = \mathbf{df} \odot \{1, \ldots, n!\}$. This is the truncated Taylor polynomial, which may converge to f [27,28]. By applying the Taylor formula for all the n given points x_i, with $i = 1 \ldots n$, we obtain $\mathbf{f} = \mathbf{Va}$, where $\mathbf{V}$is the Vandermonde matrix, with elements $v_{i,j} = x_i^{j-1}$, where $j = 1 \ldots n$ [29,30].

$$\mathbf{V} = \begin{pmatrix} 1 & x_1 & x_1^2 & \ldots & x_1^{n-1} \\ 1 & x_2 & x_2^2 & \ldots & x_2^{n-1} \\ 1 & x_3 & x_3^2 & \ldots & x_3^{n-1} \\ \vdots & \vdots & \vdots & \ddots & \vdots \\ 1 & x_n & x_n^2 & \ldots & x_n^{n-1} \end{pmatrix}$$

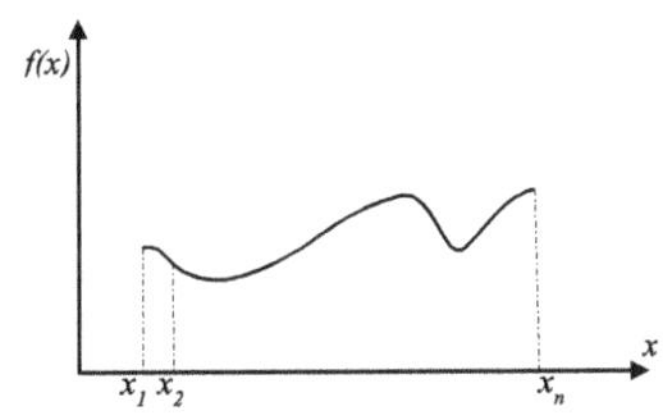

Figure 1. Given values of $f(x)$ at points x_i for the approximation of f by inverting the corresponding Vandermonde matrix $\mathbf{V}$.

The square Vandermonde matrix for distinct x_i is invertible, with $\det(\mathbf{V}) = \prod_{1 \leq i < j \leq n} (x_j - x_i)$ [31] and inverse matrix $\mathbf{V^{-1}}$=$\mathbf{U^{-1}L^{-1}}$, where the elements l_{ij} of $\mathbf{L^{-1}}$, and u_{ij} of $\mathbf{U^{-1}}$, are given by

$l_{ij} = \left\{\prod_{k=1(k\neq j)}^{i} \frac{1}{x_j - x_k}; 0 \forall i < j; l_{11} = 1\right\}$, and $u_{ij} = \{u_{i-1,j-1} - u_{i,j-1}x_{j-1}; u_{i1} = 0; u_{ii} = 1, u_{oj} = 0\}$ [32].

Hence, we have closed-form formulas for the matrix $\mathbf{V}^{-1}$ and for $\det(\mathbf{V})$, which is later used for the comparison among the various digits utilized in the calculations. Accordingly, we can compute the polynomial factors $\mathbf{a} = \{a_1, a_2, \ldots, a_n\}$ by using the following:

$$\mathbf{a} = \mathbf{V}^{-1}\mathbf{f},$$

We can also compute the corresponding errors:

$$\mathbf{e} = \mathbf{Va} - \mathbf{f}.$$

Some errors $\mathbf{e}$ are inevitable due to the truncation of the Taylor series, which theoretically comprise infinite terms, to Taylor polynomials that utilize a number of terms n. The computation of $\mathbf{a}$ with floating point arithmetic exhibits significant errors $\mathbf{e}$ in the inversion as well as the determinant calculation, with respect to their theoretical values from the closed-form formulas and numerical values computed by a machine.

3. Function Approximation in HAP

We demonstrate the proposed numerical scheme in a variety of numerical methods, analytic functions, and calculation digits. We begin with some basic operations.

3.1. Basic Operations

For the simple function$f(x) = \sin(x)$, the theoretical Taylor series exhibits an alternating sign with intermediate zero coefficients

$$\sin x = \sum_{n=0}^{\infty} \frac{(-1)^n}{(2n+1)!} x^{2n+1} = 0 + x + 0 - \frac{x^3}{3!} + 0 + \frac{x^5}{5!} - \ldots,$$

Hence, according to the presented method, the factors $\mathbf{a} = \{a_1, a_2, \ldots, a_n\}$ should be equal to $\left\{0, 1, 0, -\frac{1}{3!}, 0, \frac{1}{5!}, -\ldots, \frac{1}{n!}\right\}$ for a truncated series with n terms. However, the computation of $\mathbf{V^{-1}}$, as well as the $\det(\mathbf{V})$, exhibits great variation with the calculation precision in bits p (approximately equivalent to $p/3$ digits), when computed numerically or analytically using formulas. Table 1 presents such variation for $f(x) = \sin(x)$, with $L = 1, n = 201$, $dx = 2L/(n-1) = 10^{-2}$ and $x \in [-L, L]$. The subscript "an" denotes the analytical value and "nu" the numerical one, as computed in variable precision $p = 50$ to 2000 bits.

Table 1. Variation of $\mathbf{V}^{-1}$, $\det(\mathbf{V})$, and $\mathbf{a}$, with the calculation precision in bits p, for the same example.

Error vs. Precision (p)	$p = 50$	$p = 100$	$p = 500$	$p = 1000$	$p = 2000$
$\det \mathbf{V}_{an} - \det \mathbf{V}_{nu}$	3.866×10^{-2341}	4.300×10^{-4106}	-2.735×10^{-6810}	-3.741×10^{-6960}	-1.853×10^{-7261}
$\max \lvert \mathbf{V}_{an}^{-1} - \mathbf{V}_{nu}^{-1} \rvert$	9.739×10^{100}	4.911×10^{94}	1.242×10^{38}	1.124×10^{-111}	5.504×10^{-413}
$\max \lvert \mathbf{a}_{an} - \mathbf{a}_{nu} \rvert$	4.029×10^{1}	1.813×10^{0}	9.252×10^{-18}	9.252×10^{-18}	9.252×10^{-18}

In Table 1, a high variation in the differences among $\mathbf{V^{-1}}_{an}$ and $\mathbf{V^{-1}}_{nu}$ is revealed, from $9.739 \times 10^{+100}$ for $p = 50$ bits, which is approximately equal to floating point precision, to 5.504×10^{-413} for $p = 2000$ bits. Accordingly, the maximum differences between $\mathbf{a}_{an}^{-1}$ and $\mathbf{a}_{nu}^{-1}$ are $4.029 \times 10^{+01}$ for $p = 50$ bits and 9.252×10^{-18} for $p \geq 500$ bits. It is important to underline that all the calculations are for the same example and the same approximation scheme. Apparently, the errors of $O(10^{-16})$ cannot be considered as negligible. The significance of the precise computation is further demonstrated for the corresponding differences in the calculation of the determinant, with an analytical value constant at 1.647×10^{-6754} and the corresponding differences from the computed values varying from 3.866×10^{-2341} to -1.853×10^{-7261}, with alternating signs, again for the same example. In Table 1, we also show that as the determinants' difference shortens, the same stands for the inversion errors.

Digits accuracy exhibits great variation among the computed $1/r$, as well. The precise calculation of $\mathbf{V}$ and $\mathbf{V^{-1}}$ makes the computation of $1/r$convergent, as the calculated $\limsup_{n\to\infty} \sqrt[n]{|a_n|} \simeq \liminf_{n\to\infty} \sqrt[n]{|a_n|}$. Particularly in Figure 2a, the computed $1/r$ with a high accuracy ($p = 2000$ bits) exhibits a clear convergent pattern, whereas, for a standard accuracy ($p = 50$ bits), the corresponding $1/r$ is disoriented and does not converge (Figure 2b). Similarly, for the vector $\mathbf{a}$, the maximum absolute differences among the analytical and numerical values vary between $4.029 \times 10^{+01}$ and 9.252×10^{-18}.

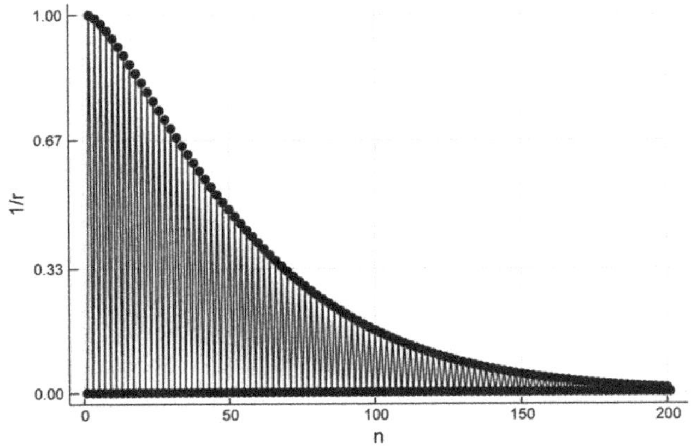

(a)

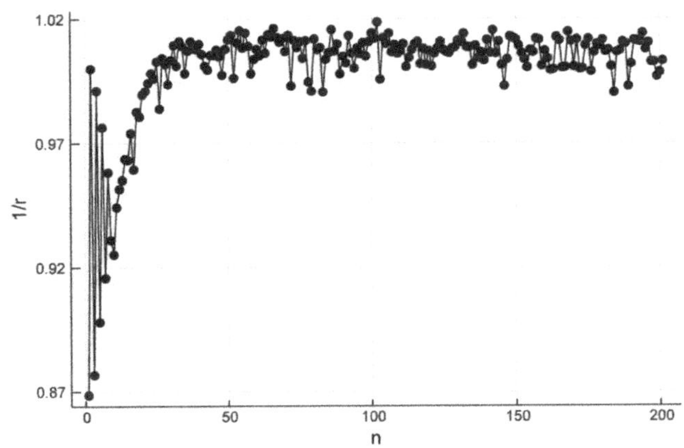

(b)

Figure 2. Radius of convergence for the computed Taylor expansion of $f(x) = \sin(x)$ for the same domain and different computational precisions. (**a**) $p = 2000$ bits (**b**) $p = 50$ bits.

3.2. Function Approximation

As $f(x) = \sin(x)$, we have that $|f^{n+1}(x)| \leq 1$; hence, the theoretical remainder of the approximation, when using $n = 200$ terms of the Taylor series, is bounded by $|R_n(x)| \leq \frac{1}{(n+1)!}|1-0|^{n+1} = 6.308 \times 10^{-378}$. In Table 2, the differences among computed and analytical values of f at x and $x_i = x + dx/2$ are presented.

Table 2. Variation of approximation errors with the calculation precision in bits p.

Error vs. Precision (p)	$p = 50$	$p = 100$	$p = 500$	$p = 1000$	$p = 2000$
$\max\|f_{an}(x) - f_{nu}(x)\|$	1.708×10^{-12}	3.045×10^{-28}	1.231×10^{-148}	3.770×10^{-299}	3.475×10^{-600}
$\max\|f_{an}(x_i) - f_{nu}(x_i)\|$	5.932×10^{-08}	2.045×10^{-15}	3.673×10^{-96}	2.373×10^{-246}	9.909×10^{-407}

Interestingly, although for $p = 50$, the approximation error for $f(x)$ on the given points x is 1.708×10^{-12}, the corresponding interpolation error on x_i is 5.932×10^{-8} (Table 2). However, the Runge phenomenon, which is severe at the boundaries, is eliminated for $p > 500$.

3.3. Extrapolation

The extrapolation problem of given data is a highly unstable process [33]. Recent results have highlighted the ability of extended spans when using a high arithmetic precision [1]. In Figure 3, the highly extended extrapolation span for $f(x) = \sin(x)$ is depicted. The extrapolation errors start becoming visible only for $x > 73L$. We should highlight that this is consistent with the corresponding theory as, for this function, the computed $1/r = \limsup_{n\to\infty} \sqrt[n]{|a_n|}$ takes the values of 0.0178, 0.0169, 0.0161, 0.0152, 0.0145, and 0.0137 for the higher values of n (Figure 2a). Accordingly, we may write that $r = 1/0.0137 \simeq 72.99$, which is equal to the observed extrapolation span. Accordingly, the extrapolation lengths for $p = 1000$ are 12.141 according to the root test $1/r$, and in the actual computations, the errors are >1 for $x > 12.150$; and similarly, for $p = 500$, the root test value is 2.154 and the computed value is 2.230, as illustrated in Figure 3. Hence, interestingly, utilizing this approach, we may predict not only determine the behavior of the approximated unknown function within the given domain, but its extrapolation spans as well, and, hence, the prediction ability.

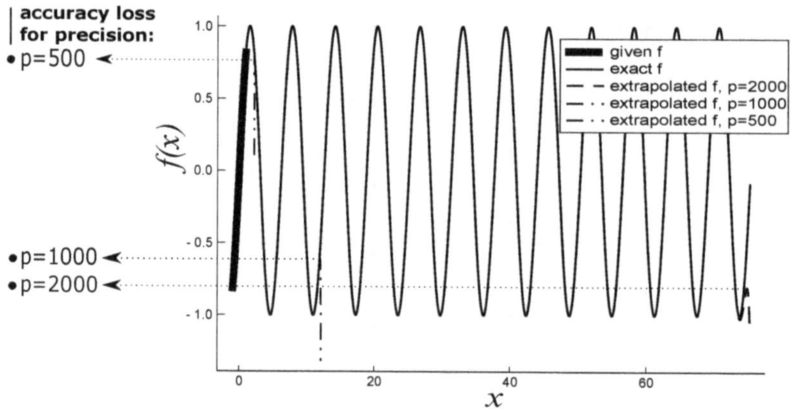

Figure 3. Extrapolation of f for varying values of arithmetic precision p. For $p = 2000$, the extrapolation errors are visible only for $x > 73L$ without any periodicity information given.

Furthermore, in Figure 4, the extrapolation of $f(x) = ln(x+1)$ (a) and $f(x) = arctan(x)$ (b) is illustrated by varying the precision p employed in the calculations. In both cases, when utilizing the standard precision $p = 50$, the extrapolation span is very short, in contrast to increased precision, such as $p = 100$, $p = 200$, and $p = 1000$.

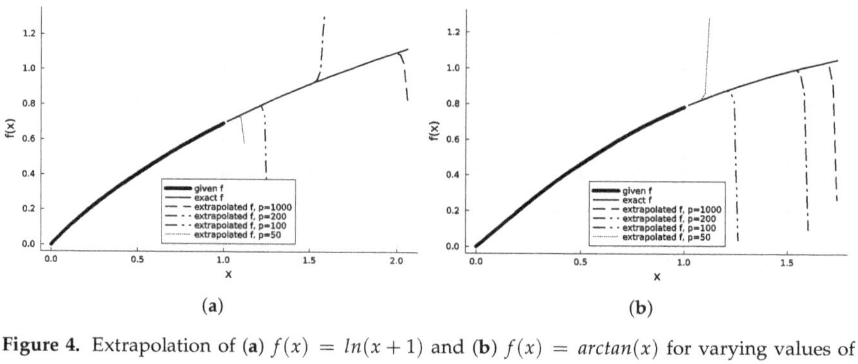

Figure 4. Extrapolation of (**a**) $f(x) = ln(x+1)$ and (**b**) $f(x) = arctan(x)$ for varying values of precision p.

3.4. Numerical Integration

We calculated the vector **a**; hence, we know an approximation of $f(x) \cong a_0 + a_1 x + a_2 x^2 + \cdots + a_n x^n$. By integrating the Taylor polynomial of f, the indefinite integral is

$$F(x) \cong a_0 x + \frac{a_1 x^2}{2} + \frac{a_2 x^3}{3} + \cdots + \frac{a_n x^{n+1}}{n+1} + c.$$

The only unknown quantity is c, which may be calculated by the supplementary constraint that $F(-L) = 0$; hence, $c \cong -a_0 L - \frac{a_1 L^2}{2} - \frac{a_2 L^3}{3} - \cdots - \frac{a_n L^{n+1}}{n+1}$. $f(x) = \sin(x)$, hence $F(x) = -\cos(x)$. The proposed scheme offers a direct computation of the integrals, as the vector **a** is known. In Table 3, the very low errors of numerical integration are demonstrated, as well as the significance of the studied digits. The numerical integration errors in Table 3 are computed as $e = F_{an} - F_{nu}$, where F_{an} is the analytical computation $F_{an} = \int_{-L}^{L} f(x)dx = -\cos(-L) + \cos(L) = 0$, for the case of $sin(x)$, and F_{nu} is the corresponding numerical computation, utilizing the computed **a**.

Table 3. Numerical integration errors.

Error vs. Precision (p)	$p = 50$	$p = 100$	$p = 500$	$p = 1000$	$p = 2000$
$F_{an} - F_{nu}$	1.502×10^{-09}	3.957×10^{-17}	1.226×10^{-97}	2.431×10^{-249}	-1.028×10^{-548}

3.5. Numerical Differentiation

The derivatives of f are directly computed by

$$\mathbf{a} = \{a_1, a_2, \ldots, a_n\} = \left\{ f(x_0), \frac{f'(x_0)}{1!}, \frac{f''(x_0)}{2!}, \ldots, \frac{f^{(n)}(x_0)}{n!} \right\} = \mathbf{df} \odot \mathbf{n}!,$$

with **df** denoting the vector of the n ordinary derivatives of f and $\mathbf{n}!$ denoting the vector of the n factorials. The $k^{th} < n$ derivative at any other point $x \neq x_0$ may easily be computed by Equation (1), deriving $f'(x) \cong 0 + a_1 + 2a_2 x + 3a_3 x^2 + \cdots + n a_n x^{n-1}$, $f''(x) \cong 0 + 0 + 2a_2 + 6a_3 x + \cdots + (n-1) n a_n x^{n-2}$, and, hence,

$$f^{(k)}(x) \cong k! a_k x^k + \cdots + \frac{n!}{(n-k)!} a_n x^{n-k}, \quad (2)$$

where the factors $\{a_k, a_{k+1}, \ldots, a_n\}$ have already been computed by **a**. We demonstrate the efficiency of the numerical differentiation in the following example apropos the solution of differential Equations.

3.6. Solution of Ordinary Differential Equations (ODEs)

The solution we investigate utilizes the constitution of the matrices representing the derivatives of each element of **V** in HAP. For example:

$$\mathbf{dV} = \begin{bmatrix} 0 & 1 & 2x_1 & \cdots & (n-1)x_1^{n-2} \\ 0 & 1 & 2x_2 & \cdots & (n-1)x_2^{n-2} \\ 0 & 1 & 2x_3 & \cdots & (n-1)x_3^{n-2} \\ \vdots & \vdots & \vdots & \ddots & \vdots \\ 0 & 1 & 2x_n & \cdots & (n-1)x_n^{n-2} \end{bmatrix}, \text{ and } \mathbf{d^2V} = \begin{bmatrix} 0 & 0 & 2 & \cdots & (n-1)(n-2)x_1^{n-3} \\ 0 & 0 & 2 & \cdots & (n-1)(n-2)x_2^{n-3} \\ 0 & 0 & 2 & \cdots & (n-1)(n-2)x_3^{n-3} \\ \vdots & \vdots & \vdots & \ddots & \vdots \\ 0 & 0 & 2 & \cdots & (n-1)(n-2)x_n^{n-3} \end{bmatrix},$$

and so on. By utilizing such matrices, we can easily constitute a system of equations representing the differential equation at points x_i. To demonstrate the unified approach to the solution of differential equations, we consider the bending of a simply supported beam [34], with the governing equation given below:

$$EI\frac{d^4w}{dx^4} = q(x) \tag{3}$$

where E is the modulus of elasticity, I the moment of inertia, w the sought solution representing the deflection of the beam, and q the external load. For $E = I = L = 1, q(x) = 0$, and fixed boundary conditions $w(0) = 0$, $\frac{dw}{dx}\Big|_{x=o} = 0, w(L) = 1/100$, $\frac{dw}{dx}\Big|_{x=L} = 0$, we may write Equation (3) supplemented by the boundary conditions in matrix form as follows:

$$\begin{bmatrix} 0 & 0 & 0 & 0 & 24 & \dots & (n-1)(n-2)(n-3)(n-4)x_1^{n-5} \\ 0 & 0 & 0 & 0 & 24 & \dots & (n-1)(n-2)(n-3)(n-4)x_2^{n-5} \\ 0 & 0 & 0 & 0 & 24 & \dots & (n-1)(n-2)(n-3)(n-4)x_3^{n-5} \\ \vdots & \vdots & \vdots & \vdots & \vdots & \ddots & \vdots \\ 0 & 0 & 0 & 0 & 24 & \dots & (n-1)(n-2)(n-3)(n-4)x_n^{n-5} \\ 1 & 0 & 0 & 0 & 0 & \cdots & 0 \\ 0 & 1 & 0 & 0 & 0 & \cdots & 0 \\ L & 0 & 0 & 0 & 0 & \cdots & 0 \\ 0 & L & 0 & 0 & 0 & \cdots & 0 \end{bmatrix} \begin{Bmatrix} a_0 \\ a_1 \\ a_2 \\ \vdots \\ a_n \end{Bmatrix} = \begin{Bmatrix} p_0 \\ p_1 \\ p_2 \\ \vdots \\ p_n \\ w_0 \\ w'_0 \\ w_L \\ w'_L \end{Bmatrix}$$

Solving for **a** and utilizing matrix **V**, we derive the sought solution using $\mathbf{w} = \mathbf{Va}$. The exact solution is

$$EIw(x) = \frac{-2EI}{L^3}x^3 + \frac{3EI}{L^2}x^2,$$

and, hence, the exact $\mathbf{a} = \{0, 0, 3, -2, 0, \dots, 0\}$. In Figure 5, the ability of a high precision ($p = 1000$) to identify the exact weights **a**is revealed, while the $p = 50$ bits accuracy fails dramatically for such identification. However, they exhibit a lower deviation than the interpolation problem, probably due to the imposition of the boundary conditions.

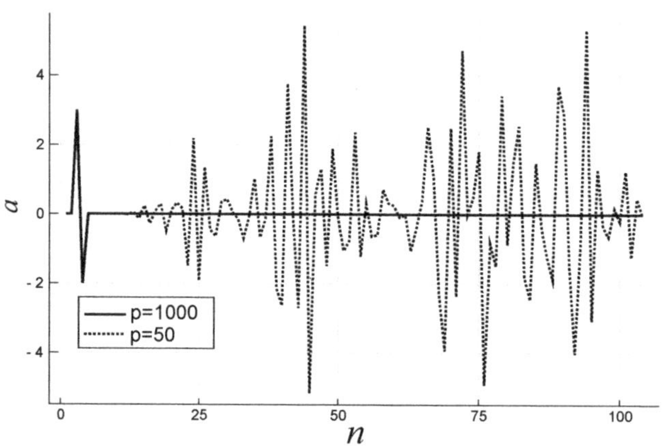

Figure 5. Calculated **a** for $p = 50$ and $p = 1000$ bits accuracies. Low arithmetic precision yields incorrect coefficients **a** for the same problem and data.

3.7. System Identification

The inverse problems, that is, the identification of the system which produced a governing differential law [35], is of great interest as this law rigorously describes the behavior of a studied system. We demonstrate the ability of high-precision Taylor polynomials to perform a rapid and precise identification of unknown systems.

Let t be an input variable and s a measured response. As presented above, we may easily compute $\mathbf{a} = \{a_1, a_2, \ldots, a_n\}$, by $\mathbf{a} = \mathbf{V}^{-1}\mathbf{s}$. Here, we assume the existence of a differential operator T, such that $T(s) = c$. According to [35], we may write T as a power series as $T(s) = \sum_{i,j,k=0}^{2} b_{ijk} s^i \dot{s}^j \ddot{s}^k = b_{000} + b_{100}s + b_{010}\dot{s} + b_{001}\ddot{s} + b_{200}s^2 + b_{110}s\dot{s} + b_{101}s\ddot{s} + b_{020}\dot{s}^2 + b_{011}\dot{s}\ddot{s} + b_{002}\ddot{s}^2$ and by setting $c' = c - b_{000}$; and assuming a linear approximation, we derive

$$1 = \frac{b_{100}s + b_{010}\dot{s} + b_{001}\ddot{s}}{c'}.$$

Applying the later for all x_i and writing the resulting system in matrix form, we obtain

$$[\mathbf{Va} + \mathbf{dVa} + \mathbf{d}^2\mathbf{Va}]\mathbf{b}^T = \{\mathbf{1}\}, \tag{4}$$

where $\{\mathbf{1}\} = \{1, 1, \ldots, 1\}$. Solving for $\mathbf{b}$, we obtain the weights of the derivatives in the differential operator $T(s)$.

For example, if we apply the previous for data of Newton's second law [30] of motion $s(t) = t^2$, with s indicating space and t time, we may calculate vectors $\mathbf{a}$ and solve Equation (4) for $\mathbf{b}$; with $c' = 1$ and a $p = 1000$ bits precision, we derive that $\mathbf{b} = \{0, 0, 1/2\} + O(10^{-270})$, and, hence, $\frac{1}{2}\ddot{s} = 1 \to \ddot{s} = 2$, which is equivalent to $\ddot{s} = a$, where $a = \frac{F}{m} = 2$, which represents the external source that produces $s(t) = t^2$.

We assume that $1 = b_{100}s + b_{010}\dot{s} + b_{001}\ddot{s}$; hence, by integrating s twice, with $S = \int s$ and $SS = \iint s$, and utilizing the interval $[0, t]$, we obtain $t + c_1 = b_{100}(S(t) - S(0)) + b_{010}(s(t) - s(0)) + b_{001}(\dot{s}(t) - \dot{s}(0))$; however, $S(0) = s(0) = \dot{s}(0) = 0$. Accordingly, we may write $t = b_{100}S(t) + b_{010}s(t) + b_{001}\dot{s}(t)$, and if we integrate for a second time in the interval $[0, t]$, we obtain

$$t^2/2 = b_{100}(SS(t) - SS(0)) + b_{010}(S(t) - S(0)) + b_{001}(s(t) - s(0)),$$

and by using $SS(0) = 0$, we obtain

$$s(t) = {}^{t^2}/_{2 - b_{100}SS(t) - b_{010}S(t)}/_{b_{001}} \tag{5}$$

The integrals of s, $\int s$, and $\iint s$ can be approximated with a high accuracy by utilizing, accordingly, the procedure discussed in Section 3.4, by using the integrals of the obtained Taylor polynomials,

$$\int s \cong a_0 t + \frac{a_1 t^2}{2} + \frac{a_2 t^3}{3} + \cdots + \frac{a_n t^{n+1}}{n+1}$$

$$\iint s \cong \frac{a_0 t^2}{2} + \frac{a_1 t^3}{6} + \frac{a_2 t^4}{12} + \cdots + \frac{a_n t^{n+2}}{(n+1)(n+2)},$$

as well as the corresponding matrices for all the given t_i,

$$\mathbf{IV} = \begin{bmatrix} 1 & 1/2 & t_1/3 & \ldots & t_1^{n+1}/(n+1) \\ 1 & 1/2 & t_2/3 & \ldots & t_2^{n+1}/(n+1) \\ 1 & 1/2 & t_3/3 & \ldots & t_3^{n+1}/(n+1) \\ \vdots & \vdots & \vdots & \ddots & \vdots \\ 1 & 1/2 & t_n/3 & \ldots & t_n^{n+1}/(n+1) \end{bmatrix}$$

$$\mathbf{IIV} = \begin{bmatrix} 1/2 & 1/6 & t_1/12 & \ldots & t_1^{n+2}/(n+1)/(n+2) \\ 1/2 & 1/6 & t_2/12 & \ldots & t_2^{n+2}/(n+1)/(n+2) \\ 1/2 & 1/6 & t_3/12 & \ldots & t_3^{n+2}/(n+1)/(n+2) \\ \vdots & \vdots & \vdots & \ddots & \vdots \\ 1/2 & 1/6 & t_n/12 & \ldots & t_n^{n+2}/(n+1)/(n+2) \end{bmatrix}.$$

The calculated impact of b_{001} for the $p = 50$ and $p = 1000$ bits accuracy is revealed by the resulting extrapolation curves beyond the observed domain, utilizing Equation (5). For the $p = 50$ bits accuracy, for given data in the domain $[0,1]$, we may extrapolate only up to a short time ($t' = 1.343$) after the last given $t_{end} = 1.000$, with threshold for errors < 1.000, while for $p = 2000$ bits, the corresponding t' attains the remarkably high value of $9.621 \times 10^{+10}$, highlighting the extrapolation power of high arithmetic precision.

4. Functions in Multiple Dimensions

4.1. Multidimensional Interpolation

The Taylor series of $f(x,y)$, depending on two variables $x, y \in \Omega$, where Ω is a closed disk about the center x_0, y_0, may be written utilizing the partial derivatives of f [31,32] in the form of $f(x,y) = f(a,b) + (x-a)f_x(a,b) + (y-b)f_y(a,b) + \frac{1}{2!}((x-a)^2 f_{xx}(a,b) + 2(x-a)(y-b)f_{xy}(a,b) + (y-b)^2 f_{yy}(a,b)) + \ldots$, which, in vector form, is written as

$$f(\mathbf{x}) = f(\mathbf{x}_0) + (\mathbf{x}-\mathbf{x}_0)^T Df(\mathbf{x}_0) + \frac{1}{2!}(\mathbf{x}-\mathbf{x}_0)^T \left\{D^2 f(\mathbf{x}_0)\right\}(\mathbf{x}-\mathbf{x}_0) + \cdots,$$

where $D^2 f(\mathbf{x}_0)$ is the Hessian matrix at $\mathbf{x}_0$.

Let n be the number of given points of $f(x_i, y_j)$, with $i, j \in (1, 2, \ldots, n)$. In order to constitute the approximating polynomial of $f(x,y)$ with high-order terms and formulate the $\mathbf{V}$ matrix with dimensions $n \times n$, we consider all possible combinations of $\{n_i, n_j \in (0, 1, \ldots, n-1) \mid n_i + n_j \leq n-1\}$. Hence, we may write the following for all the given x_i:

$$\mathbf{V}(\mathbf{x}_i,\mathbf{y}_j) = \begin{bmatrix} 1 & x_1 & y_1 & x_1 y_1 & x_1^2 & y_1^2 & \ldots & x_1^{n_k} y_1^{n_l} \\ 1 & x_2 & y_2 & x_2 y_2 & x_2^2 & y_2^2 & \ldots & x_2^{n_k} y_2^{n_l} \\ \ldots & \ldots & \ldots & \ldots & \ldots & \ldots & \ddots & \ldots \\ 1 & x_n & y_n & x_n y_n & x_n^2 & y_n^2 & \ldots & x_n^{n_k} y_n^{n_l} \end{bmatrix},$$

with $k + l = n - 1$. Thus, we can approximate f with n polynomial terms using the following:

$$\mathbf{f} = \mathbf{V}\mathbf{a} \rightarrow \mathbf{a} = \mathbf{V}^{-1}\mathbf{f} \tag{6}$$

The computation of $\mathbf{a}$ with Equation (6) permits the computation of $f(\widehat{x}_i, \widehat{y}_j)$, for any $\widehat{x}_i, \widehat{y}_j \in \Omega$, by utilizing the corresponding $\widehat{\mathbf{V}}$.

Let $f(x,y) = \sin(5x) + \cos(e^{2y})$. We approximate f with $n = 300$ random values $x_i, y_i \in [-0.5, 0.5]$, and, later, we interpolate f with $n = 300$ random values $\widehat{x}_i, \widehat{y}_{j_i} \in [-0.35, 0.35]$. In Figure 6, the exact and approximated values $f(\widehat{x}_i, \widehat{y}_j)$ are depicted for $p = 2000$ and $p = 50$ bits accuracies. Apparently, for the same interpolation problem formulation in three dimensions, the computational precision p dramatically affects the results. The

$$\max\left| f(x_i, y_j)_{analytical} - f(\widehat{x}_i, \widehat{y}_j)_{numerical} \right|$$

equals 8.570×10^{-09} for $p = 2000$ and $1.286 \times 10^{+01}$ for $p = 50$ bits. The polynomials' weights $\mathbf{a}$ were calculated by first computing $\mathbf{V}^{-1}$ by solving $\mathbf{V} \backslash \mathbf{I}$; hence, $\mathbf{a} = \mathbf{V}^{-1}\mathbf{f}$ because $\mathbf{a} = \mathbf{V} \backslash \mathbf{I}$ exhibits significant errors. The calculation of the inverse of generic matrices, as well as the solution of systems of Equations in a high precision, is a topic for future research.

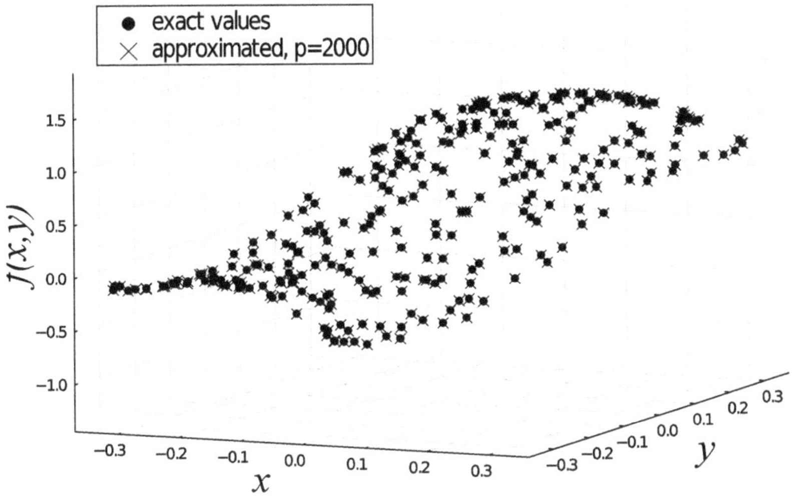

(a)

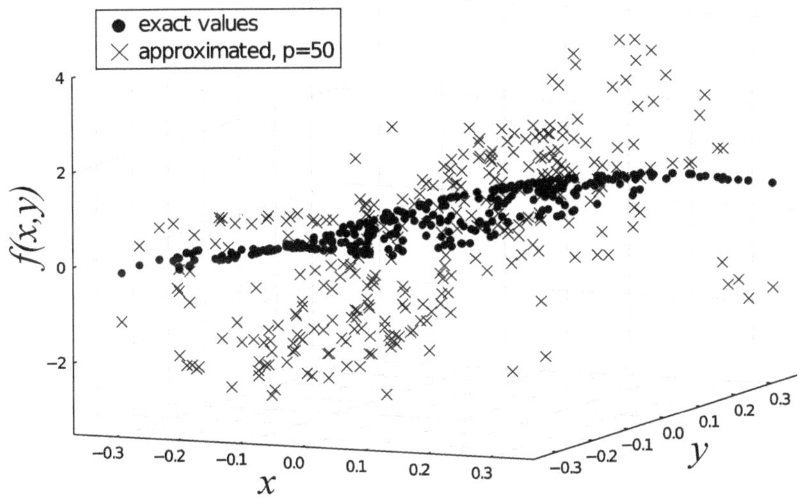

(b)

Figure 6. Exact and approximated values of f for precision $p = 2000$ bits (**a**) and $p = 50$ bits (**b**). We can observe that by utilizing enough digits, we have a precise approximation in contrast to a standard precision, indicating a computational deficiency and not a methodological one.

4.2. Solution of Partial Differential Equations

We present the ability of a high precision to solve partial differential equations by considering a plate without axial deformations and vertical load $q(x, y)$. The governing Equation [36] has the following form:

$$\frac{\partial^4 w}{\partial x^4} + 2\frac{\partial^4 w}{\partial x^2 \partial y^2} + \frac{\partial^4 w}{\partial y^4} = -\frac{q}{D} \tag{7}$$

that is, $\nabla^2\nabla^2 w = -\frac{q}{D}$, where $D := \frac{Eh^3}{12(1-v^2)}$, E is the modulus of elasticity, v is the Poisson constant, and h is the slab's height.

The sought solution $w(x,y)$ is the slab's deformation within the boundary conditions $w_b(\mathbf{x}_b, \mathbf{y}_b)$ along some boundaries $b = \{1,2,\ldots\}$. In order to solve Equation (7), we approximate

$$\mathbf{w} = \mathbf{V}\mathbf{a}$$

using the approximation scheme of Equation (6), and as the vector $\mathbf{a}$ is constant, we obtain $\mathbf{w}_{x^4}=\mathbf{V}_{x^4}\mathbf{a}$, $\mathbf{w}_{y^4}=\mathbf{V}_{y^4}\mathbf{a}$, and $\mathbf{w}_{x^2y^2}=\mathbf{V}_{x^2y^2}\mathbf{a}$, with $\mathbf{w}_{x^ky^l}$ denoting the partial derivative of w of order k over x and l over y, $\frac{\partial^{k+l}w}{\partial x^k \partial y^l}$, for all given x_i, y_j with $i,j \in (1,2,\ldots,n)$. Utilizing this notation, we may write Equation (7) for all x_i, y_j in matrix form as

$$\left[\mathbf{V}_{x^4} + 2\mathbf{V}_{x^2y^2} + \mathbf{V}_{y^4}\right]\mathbf{a} = \mathbf{q}.$$

By applying some boundary conditions, we may write for the same $\mathbf{a}$,

$$\begin{bmatrix} \mathbf{V}_{x^4} + 2\mathbf{V}_{x^2y^2} + \mathbf{V}_{y^4} \\ \mathbf{V}(x_1,y_1) \\ \mathbf{V}_x(x_2,y_2) \\ \cdots \end{bmatrix} \times \mathbf{a} = \begin{bmatrix} \mathbf{q} \\ w(x_1,y_1) \\ \frac{\partial w}{\partial x}\Big|_{(x_1,y_1)} \\ \cdots \end{bmatrix} \rightarrow$$

$$\mathbf{a} = \begin{bmatrix} \mathbf{V}_{x^4} + 2\mathbf{V}_{x^2y^2} + \mathbf{V}_{y^4} \\ \mathbf{V}(x_1,y_1) \\ \mathbf{V}_x(x_2,y_2) \\ \cdots \end{bmatrix}^{-1} \times \begin{bmatrix} \mathbf{q} \\ w(x_1,y_1) \\ \frac{\partial w}{\partial x}\Big|_{(x_1,y_1)} \\ \cdots \end{bmatrix}. \tag{8}$$

By computing $\mathbf{a}$, we then obtain the sought solution as $\mathbf{w} = \mathbf{V}\mathbf{a}$.

For example, for a simply supported slab, the boundary conditions are $w(x_b, y_b) = w_b$ for some boundary b. We consider a square slab, with $n = 20$ divisions per dimension, $dx = 1/99$, $L = (n-1)dx$, and $w(x_b, y_b) = 0$, at the four linear boundaries, and $\mathbf{q} = \mathbf{1}$, the normalized load to comprise values of 1 everywhere (Equation (7)). After the computation of $\mathbf{a}$ with Equation (8), we may easily compute the corresponding shear forces, which are defined by

$$Q_x = -D\frac{\partial}{\partial x}\left(\frac{\partial^2 w}{\partial x^2} + \frac{\partial^2 w}{\partial y^2}\right), Q_y = -D\frac{\partial}{\partial y}\left(\frac{\partial^2 w}{\partial x^2} + \frac{\partial^2 w}{\partial y^2}\right).$$

We utilize the computed $\mathbf{a}$ and matrices $\mathbf{V}_{xxx}, \mathbf{V}_{xyy}, \mathbf{V}_{yxx}, \mathbf{V}_{yyy}$. Newton's equilibrium states that the total shear force at the boundaries should be equal to the total applied force. For a constant load over the plate, the Equilibrium error

$$\max\left|\int_A q(x,y) - \sum Q_{x,y}\right|,$$

for $p = 50$ bits is 6.924×10^{-05}, and for $p = 2000$, it is 2.242×10^{-591}. We observe that there is a large difference, though the errors are small, even with $p = 50$ bits. Interestingly, utilizing a concentrated load by loading the nodes close to $(0,0)$, the inversion error

$$\max\left|\begin{bmatrix} \mathbf{V}_{x^4} + 2\mathbf{V}_{x^2y^2} + \mathbf{V}_{y^4} \\ \mathbf{V}(x_1,y_1) \\ \mathbf{V}_x(x_2,y_2) \\ \cdots \end{bmatrix}^{-1} \times \begin{bmatrix} \mathbf{V}_{x^4} + 2\mathbf{V}_{x^2y^2} + \mathbf{V}_{y^4} \\ \mathbf{V}(x_1,y_1) \\ \mathbf{V}_x(x_2,y_2) \\ \cdots \end{bmatrix} - \mathbf{I}\right|,$$

for $p = 50$ bits is 43.988, and for $p = 2000$, it is 4.381×10^{-587}, further highlighting the significance of accuracy in the calculations.

5. Conclusions

Function approximation exists in the core calculations of computational mechanics, with implications for other disciplines. In this work, a high arithmetic precision, when applied to Taylor polynomials, is found capable of executing various numerical tasks precisely. Particularly, a high arithmetic precision significantly improves accuracy in solving beam deflection equations, demonstrating the importance of computational precision in the solution of ODEs. A high precision significantly enhances the solution accuracy of partial differential equations for slab deformation under a vertical load, highlighting the critical role of computational precision in PDEs. Furthermore, traditional issues like the Runge phenomenon, commonly encountered in numerical approximations, are eliminated with the use of a HAP. The radius of convergence for the Taylor series is precisely computable using a HAP, providing valuable insights into the interpolation accuracy and potential extrapolation range of an unknown function.

Overall, the use of Taylor polynomials in a high arithmetic precision showcases potential as a unified approach to various numerical computations, delivering highly accurate results and revealing that some numerical instabilities are due to computational inaccuracies rather than methodological issues. Future research can include parallel computing techniques or optimized matrix inversion strategies to deal with the Vandermonde matrix and other related computational challenges in HAP. Taylor polynomials with a high precision could also be applied to more complex systems and geometries in computational mechanics, as well as other engineering problems involving function approximation, such as fluid dynamics and quantum physics. Extending the research to high-dimensional problems where function approximation becomes significantly more complicated could also be an important field, addressing the practical aspects of high-precision calculations for partial differential equations and integral equations in a high-dimensional space. The study of precision in calculations illustrates the odd but fundamental epistemological principle that even $1 + 1 = 2$ might be falsified [37].

Funding: We acknowledge support from project "SimEA" funded by the European Union's Horizon 2020 research and innovation program under Grant Agreement No. 810660.

Data Availability Statement: All the data and results may be reproduced by the computer code on GitHub https://github.com/nbakas/TaylorBigF.jl. The code is in generic form, so as to solve for any numerical problem with the discussed methods. The code is written in Julia [24], utilizing the MPFR [38] and GMP [9] Libraries.

Conflicts of Interest: The author declares no conflict of interest.

Nomenclature

x	Variable x, corresponding to $f(x)$
x_0	Initial point in the approximation
n	Number of terms in the Taylor series, also number of nodes
L	Length of the given domain
E	Modulus of elasticity
I	Inertia of the beam
$f(x)$	Analytic function
$\mathbf{f}$	Vector of function values
$\mathbf{x}$	Vector of points
r	Radius of convergence
$\mathbf{V}$	Vandermonde matrix
$\mathbf{df}$	Vector of the derivatives of the function $f(x)$
D	Flexural rigidity of plate
w	Deflection of the beam/plate
q	External load of beam/plate
$\mathbf{a}$	Coefficient vector for Taylor polynomials
v	Poisson constant
h	Slab's thickness

References

1. Bakas, N.P. Numerical Solution for the Extrapolation Problem of Analytic Functions. *Research* **2019**, *2019*, 3903187. [CrossRef] [PubMed]
2. Bailey, D.H.; Jeyabalan, K.; Li, X.S. A comparison of three high-precision quadrature schemes. *Exp. Math.* **2005**, *14*, 317–329. [CrossRef]
3. Cheng, A.H. Multiquadric and its shape parameter—A numerical investigation of error estimate, condition number, and round-off error by arbitrary precision computation. *Eng. Anal. Bound. Elem.* **2012**, *36*, 220–239. [CrossRef]
4. Huang, C.S.; Lee, C.F.; Cheng, A.H. Error estimate, optimal shape factor, and high precision computation of multiquadric collocation method. *Eng. Anal. Bound. Elem.* **2007**, *31*, 614–623. [CrossRef]
5. Sharma, N.N.; Jain, R.; Pokkuluri, M.M.; Patkar, S.B.; Leupers, R.; Nikhil, R.S.; Merchant, F. CLARINET: A quire-enabled RISC-V-based framework for posit arithmetic empiricism. *J. Syst. Archit.* **2023**, *135*, 102801. [CrossRef]
6. Lei, X.; Gu, T.; Xu, X. ddRingAllreduce: A high-precision RingAllreduce algorithm. *CCF Trans. High Perform. Comput.* **2023**, *5*, 245–257. [CrossRef]
7. Wu, C.; Xia, Y.; Xu, Z.; Liu, L.; Tang, X.; Chen, Q.; Xu, F. Mathematical modelling for high precision ray tracing in optical design. *Appl. Math. Model.* **2024**, *128*, 103–122. [CrossRef]
8. Friebel, K.F.A.; Bi, J.; Castrillon, J. Base2: An IR for Binary Numeral Types. In Proceedings of the 13th International Symposium on Highly Efficient Accelerators and Reconfigurable Technologies, Kusatsu, Japan, 14–16 June 2023; pp. 19–26. [CrossRef]
9. Granlund, T. The GNU Multiple Precision Arithmetic Library. *Free Softw. Found.* Available online: https://gmplib.org/ (accessed on 13 January 2024).
10. Amato, G.; Scozzari, F. JGMP: Java bindings and wrappers for the GMP library. *SoftwareX* **2023**, *23*, 101428. [CrossRef]
11. Guessab, A.; Nouisser, O.; Schmeisser, G. Multivariate approximation by a combination of modified Taylor polynomials. *J. Comput. Appl. Math.* **2006**, *196*, 162–179. [CrossRef]
12. Kalantari, B. Generalization of Taylor's theorem and Newton's method via a new family of determinantal interpolation formulas and its applications. *J. Comput. Appl. Math.* **2000**, *126*, 287–318. [CrossRef]
13. Berz, M.; Makino, K. Verified Integration of ODEs and Flows Using Differential Algebraic Methods on High-Order Taylor Models. *Reliab. Comput.* **1998**, *10*, 361–369. [CrossRef]
14. Yalçinbaş, S.; Sezer, M. The approximate solution of high-order linear Volterra-Fredholm integro-differential equations in terms of Taylor polynomials. *Appl. Math. Comput.* **2000**, *112*, 291–308. [CrossRef]
15. Ranjan, R.; Prasad, H.S. A novel approach for the numerical approximation to the solution of singularly perturbed differential-difference equations with small shifts. *J. Appl. Math. Comput.* **2021**, *65*, 403–427. [CrossRef]
16. Platte, R.B.; Trefethen, L.N.; Kuijlaars, A.B.J. Impossibility of Fast Stable Approximation of Analytic Functions from Equispaced Samples. *SIAM Rev.* **2011**, *53*, 308–318. [CrossRef]
17. Boyd, J.P. Defeating the Runge phenomenon for equispaced polynomial interpolation via Tikhonov regularization. *Appl. Math. Lett.* **1992**, *5*, 57–59. [CrossRef]
18. Zhang, S.Q.; Fu, C.H.; Zhao, X.D. Study of regional geomagnetic model of Fujian and adjacent areas based on 3D Taylor Polynomial model. *Acta Geophys. Sin.* **2016**, *59*, 1948–1956. [CrossRef]
19. Boyd, J.P.; Xu, F. Divergence (Runge Phenomenon) for least-squares polynomial approximation on an equispaced grid and Mock-Chebyshev subset interpolation. *Appl. Math. Comput.* **2009**, *210*, 158–168. [CrossRef]
20. Boyd, J.P.; Ong, J.R. Exponentially-convergent strategies for defeating the runge phenomenon for the approximation of non-periodic functions, part I: Single-interval schemes. *Commun. Comput. Phys.* **2009**, *5*, 484–497.
21. Taylor, B. *Principles of Linear Perspective*; Knaplock, R., Ed.; British Library: London, UK, 1715.
22. Babouskos, N.G.; Katsikadelis, J.T. Optimum design of thin plates via frequency optimization using BEM. *Arch. Appl. Mech.* **2015**, *85*, 1175–1190. [CrossRef]
23. Yiotis, A.J.; Katsikadelis, J.T. Buckling of cylindrical shell panels: A MAEM solution. *Arch. Appl. Mech.* **2015**, *85*, 1545–1557. [CrossRef]
24. Bezanson, J.; Edelman, A.; Karpinski, S.; Shah, V.B. Julia: A fresh approach to numerical computing. *SIAM Rev.* **2017**, *59*, 65–98. [CrossRef]
25. Apostol, T.M. *Calculus*; John Wiley & Sons: Hoboken, NJ, USA, 1967.
26. Browder, A. *Mathematical Analysis: An Introduction*; Springer Science & Business Media: Berlin/Heidelberg, Germany, 2012.
27. Katsoprinakis, E.S.; Nestoridis, V.N. Partial sums of Taylor series on a circle. *Ann. L'Institut Fourier* **2011**, *39*, 715–736. [CrossRef]
28. Nestoridis, V. Universal Taylor series. *Ann. de L'Institut Fourier* **2011**, *46*, 1293–1306. [CrossRef]
29. Press, W.H.; Teukolsky, S.A. *VWT, and FBP, Numerical Recipes: The Art of Scientific Computing*; Cambridge University Press: Cambridge, UK, 2007.
30. Horn, R.A.; Johnson, C.R. *Topics in Matrix Analysis*; Cambridge University Press: Cambridge, UK, 1991.
31. Ycart, B. A case of mathematical eponymy: The Vandermonde determinant. *arXiv* **2012**, arXiv:1204.4716.
32. Turner, L.R. *Inverse of the Vandermonde Matrix with Applications*; NASA–TN D-3547; NASA: Washington, DC, USA, 1966.
33. Demanet, L.; Townsend, A. Stable extrapolation of analytic functions. *Found. Comput. Math.* **2019**, *19*, 297–331. [CrossRef]
34. Boresi, A.P.; Sidebottom, O.M.; Saunders, H. Advanced Mechanics of Materials (4th Ed.). *J. Vib. Acoust. Stress Reliab. Des.* **1988**, *110*, 256–257. [CrossRef]

35. Katsikadelis, J.T. System identification by the analog equation method. *WIT Trans. Model. Simul.* **1995**, *10*, 12.
36. Katsikadelis, J.T. *The Boundary Element Method for Plate Analysis*; Elsevier: Amsterdam, The Netherlands, 2014.
37. Gregory, F.H. Arithmetic and Reality: A Development of Popper's Ideas. *Philos. Math. Educ. J.* **2011**, *26*.
38. Fousse, L.; Hanrot, G.; Lefèvre, V.; Pélissier, P.; Zimmermann, P. MPFR: A multiple-precision binary floating-point library with correct rounding. *ACM Trans. Math. Softw.* **2007**, *33*, 2. [CrossRef]

computation

MDPI

Article

Multi-Directional Functionally Graded Sandwich Plates: Buckling and Free Vibration Analysis with Refined Plate Models under Various Boundary Conditions

Lazreg Hadji [1,*], Vagelis Plevris [2], Royal Madan [3] and Hassen Ait Atmane [4,5]

1 Department of Civil Engineering, University of Tiaret, BP 78 Zaaroura, Tiaret 14000, Algeria
2 Department of Civil and Environmental Engineering, Qatar University, Doha P.O. Box 2713, Qatar; vplevris@qu.edu.qa
3 Department of Mechanical Engineering, Graphic Era (Deemed to be University), Dehradun 248002, India; royalmadan6293@gmail.com
4 Civil Engineering Department, University of Hassiba Ben Bouali, Chlef 02180, Algeria; aitatmane2000@yahoo.fr
5 Laboratory of Structures, Geotechnics and Risks, University of Hassiba Ben Bouali, Chlef 02180, Algeria
* Correspondence: lazreg.hadji@univ-tiaret.dz

Abstract: This study conducts buckling and free vibration analyses of multi-directional functionally graded sandwich plates subjected to various boundary conditions. Two scenarios are considered: a functionally graded (FG) skin with a homogeneous hard core, and an FG skin with a homogeneous soft core. Utilizing refined plate models, which incorporate a parabolic distribution of transverse shear stresses while ensuring zero shear stresses on both the upper and lower surfaces, equations of motion are derived using Hamilton's principle. Analytical solutions for the buckling and free vibration analyses of multi-directional FG sandwich plates under diverse boundary conditions are developed and presented. The obtained results are validated against the existing literature for both the buckling and free vibration analyses. The composition of metal–ceramic-based FG materials varies longitudinally and transversely, following a power law. Various types of sandwich plates are considered, accounting for plate symmetry and layer thicknesses. This investigation explores the influence of several parameters on buckling and free vibration behaviors.

Keywords: buckling; free vibration; hard core; soft core; multi-directional FGM

1. Introduction

Composite materials blend two or more substances with varying properties to create unique characteristics absent in their individual components. Functionally graded materials (FGMs) represent a specific class of composites distinguished by directional variations in their material properties [1]. While unidirectional FGMs vary their properties along a single axis, multi-directional (MD) graded materials introduce variations along multiple axes, enhancing their performance [2].

Various methodologies have been proposed for analyzing the free vibration of plates, beams, and shells, including the energy method for plates [3], Galerkin–Vlasov's method for tapered plates [4], and the Rayleigh–Ritz method for rotating hard-coating cylindrical shells [5]. For 2-D FGMs, studies have explored their free vibration behavior under different boundary conditions, revealing variations in their frequency based on the boundary conditions [6]. A semi-analytical numerical method was employed in solving the problems of the bending analysis of 2-D functionally graded (FG) circular and annular plates [7] and their impact analysis [8].

Buckling and frequency analyses were conducted on a two-directional (2-D) FG circular plate using the differential quadrature method (DQM) for both clamped and simply supported boundary conditions. The results indicated a higher critical buckling load for the clamped plate compared to the simply supported configuration [9]. Additionally, finite element analysis and analytical techniques like the third-order shear deformation plate theory have provided us with understanding of the static deflection and buckling characteristics of FGM plates [10,11].

The free vibration and buckling of 2-D FGM plates were investigated using a non-uniform rational B-spline technique [12]. Estimating the material properties of FGMs is crucial, with empirical relationships available for ideal FGMs. However, certain parameters like the stress–strain transfer ratio (q) remain mathematically unquantifiable, necessitating experimental validation [13]. Several material combination experiments have been conducted to explore the parameter q empirically [14].

A free vibration analysis of multi-directional FG piezoelectric annular plates has been conducted using the differential quadrature method (DQM). Radial and thickness gradations were incorporated, revealing that radial gradation enhances plate stiffness, resulting in higher frequency responses [15]. Static and dynamic analyses of three-dimensional shells composed of multi-directional FG material (MD FGM) have been carried out using polyhedral finite element methods. That study explored the effects of shell thickness and slenderness on structural behavior [16]. The bending analysis of multilayer panels of FGMs was performed using a higher-order layer-wise model, with Young's modulus determined via the Halpin–Tsai method and Poisson's ratio via the rule of mixture [17]. In a thermal environment, numerical and experimental investigations were used to examine the free vibration of unidirectional and bi-directional porous FG curved panels. The evaluation of temperature-dependent (TD) material properties showed lower frequencies compared to temperature-independent (TID) properties [18]. A strain gradient elasticity theory was employed to study the dynamic response of square microplates with multi-directional FGM properties under a moving concentrated load [19]. The optimization of multi-directional FG plates under thermal effects was achieved through free vibration analysis [20]. Furthermore, a free vibration analysis of tri-directional FG beams under magneto-electro-elastic fields utilized the DQM and higher-order deformation theory [21,22]. In another study, the analysis of MD FG sandwich plates covered both FG skin with a homogeneous core and FG cores with homogeneous skin configurations, with the former demonstrating higher natural frequencies [23].

Singh and Kumari [24] proposed an approximate analytical solution for analyzing the free vibration of composite FG rectangular plates. By applying a modified version of Hamilton's principle, they derived governing equations, considering all stresses and displacements as primary variables. The solution was obtained using the extended Kantorovich method, along with Fourier and power series approaches. Singh et al. [25] introduced a framework for accurately analyzing the free vibration of in-plane FG orthotropic rectangular plates integrated with piezoelectric sensory layers, considering both their elastic and viscoelastic properties. Numerical studies have explored the effects of in-plane gradation and viscoelasticity on vibration responses, revealing significant alterations in flexural frequencies and mode shapes. Vaishali et al. [26] proposed an innovative multi-physical probabilistic vibration analysis approach for FG materials. They combined Gaussian Process Regression (GPR) with finite element simulations, aided by a Monte Carlo Simulation, resulting in significant computational efficiency. By integrating machine learning with physics-based modeling, system uncertainty can be efficiently quantified.

Malikan and Eremeyev [27] developed a novel hyperbolic, polynomial higher-order elasticity theory for thick FGM beams. Their model addressed a critical drawback in material composition and incorporated a unique shape function for shear stress distribution. Through rigorous validation and comparative analyses, they demonstrated the efficacy of their approach. Their findings underscored the significance of higher-order beam theories and stretching effects. Importantly, their investigation into FGM beams with different boundary conditions revealed the marked effects of material imperfections, emphasizing the practical implications of their work for structural mechanics and material engineering.

Functionally graded materials undergo changes in their properties through adjustments in their microstructure, material composition, and porosity. Manufacturing techniques such as solid-state, liquid-state, or deposition processes enable the fabrication of these materials [28,29]. The research on FG structures, including plates, disks, and beams, has predominantly focused on unidirectional FGMs, as evidenced by the existing literature [30–32]. Fabrication methods must minimize the delamination that results from differences in the materials' physical and chemical characteristics, necessitating techniques that ensure consistent thermo-mechanical properties [33]. While existing methods suit the fabrication of unidirectional FGMs, multi-directional gradation poses unique challenges. Functionally graded additive manufacturing (FGAM) emerges as a promising solution due to its ability to produce components with material gradients in various directions, offering advantages such as reduced material wastage, the absence of tooling requirements, and decreased manufacturing time and costs [34]. Various material modeling techniques exist, with rule of mixture (ROM) models often providing the best results for material combinations when compared to experimental data [35].

A delamination analysis within multilayered FG beam configurations was performed with a specific emphasis on understanding the time-dependent strain energy release rate. The study included different models of nonlinear creep behavior, particularly in tension and compression scenarios [36]. Dastjerdi et al. [37] employed a highly efficient quasi-3D theory to investigate the nonlinear hygro-thermo-mechanical bending analysis of a thick FGM rotating disk in a hygro-thermal environment, taking porosity into account as a structural defect. Their analysis incorporates two applied quasi-3D displacement fields, where the strain along the thickness is non-zero, unlike in conventional plate theories. Karami and Ghayesh [38] explored the significance of micromechanical models in analyzing the forced vibrations of multi-layered microplates subjected to a moving load. Their microplate comprised an FGM core and metal foam face sheets. The problem was modelled using a quasi-3D shear deformable method and modified couple stress theory. The same authors [39] investigated the vibrations of sandwich microshells featuring porous FG face sheets, considering in-surface curvilinear motions. The motion equations were derived using Hamilton's principle, employing a curvilinear framework for a modified couple stress scheme that incorporated length-scale parameters. The vibration modes for curvilinear and normal displacements were assumed using trigonometric functions, and natural frequencies were determined numerically.

Despite the extensive research on unidirectional graded plates, research on multi-directional gradation remains limited. Multi-directional gradation promises optimized structures with enhanced performance compared to unidirectional grading. Thus, this paper investigates the free vibration and buckling analysis of multi-directional FG plates using refined plate theories. Our analysis considers a face sheet made of FGM with a homogeneous core composed of ceramic (hard core) or metal (soft core). The results of the proposed methods are validated against the existing literature for both buckling and free vibration analyses. Furthermore, a detailed parametric analysis explores the effects of grading index and geometry on the frequency and buckling load of multi-directional FG plates.

2. Mathematical Modeling

2.1. Preliminary Concepts and Definitions

Figure 1 depicts a multi-directional FGM sandwich plate with its dimensions along the x, y, and z axes denoted as length (a), width (b), and thickness (h), respectively. The sandwich plate comprises two face sheets (top and bottom) sandwiching a core layer. In Figure 1, the x and y axes represent the midplane, while the z axis is perpendicular to the midplane.

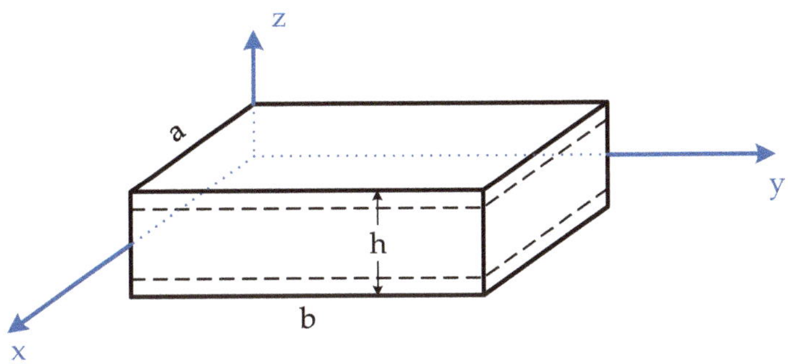

Figure 1. Multi-directional FGM sandwich plate.

A power law variation in the volume fraction of the FGM's metal and ceramic constituents is represented by Equations (1) and (2). This investigation focuses on metal (Al) and ceramic (Al_2O_3), with their compositions varying longitudinally (p_x) and transversely (p_z). The material properties of the FGM sandwich plate are influenced by the volume fractions of these constituents along both its longitudinal and transverse directions, following a power law relationship. The volume fraction of metal in a multi-directional sandwich plate is expressed as shown in Equation (1):

$$\begin{aligned} V^{(1)}(x,z) &= \left(\tfrac{z-h_1}{h_1-h_2}\right)^{p_z}\left(1-\tfrac{x}{2a}\right)^{p_x} \\ V^{(2)}(x,z) &= 1 \\ V^{(3)}(x,z) &= \left(\tfrac{z-h_4}{h_3-h_4}\right)^{p_z}\left(1-\tfrac{x}{2a}\right)^{p_x} \end{aligned} \quad (1)$$

where $V^{(n)}$ (n = 1, 2, 3) represents the volume fraction function of Layer n, while p_z and p_x denote the volume fraction indices in the transverse and longitudinal directions, respectively. Type A has an FG face sheet and homogeneous hard core, while Type B has an FG face sheet and homogeneous soft core. Figure 2 presents the different layers of the material. The layer thicknesses, denoted by the coordinate points $h_1 = -h/2$, h_2, h_3, and $h_4 = h/2$ in the z direction, determine the sandwich configuration.

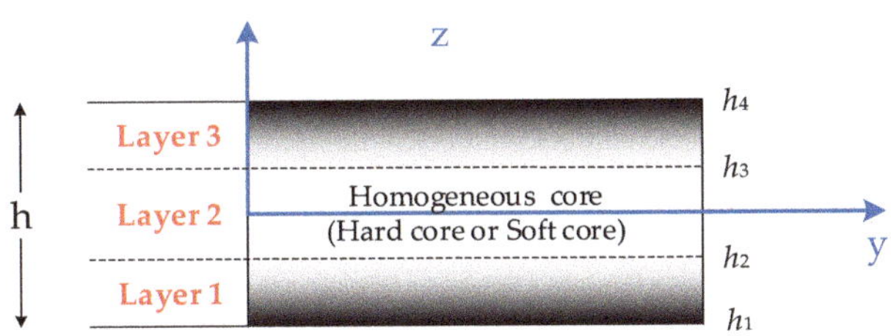

Figure 2. Material variation with thickness of multi-directional FGM sandwich plate: FGM containing face sheets and a homogeneous core (Type A: hard core, Type B: soft core).

2.2. Modeling of FG Sandwich Plate

The effective material properties of the plate, i.e., its Young's modulus E, Poisson's ratio ν, and mass density ρ, can be expressed by the rule of mixture, as shown in Equation (2) [40].

$$P^{(n)}(x,z) = (P_1 - P_2)V^{(n)}(x,z) + P_2 \quad (2)$$

where $P^{(n)}$ is the effective material property of the FGM of Layer n. For Type A, P_1 and P_2 are the properties of the top and bottom faces of Layer 1, respectively, and vice versa for Layer 3, depending on the volume fraction $V^{(n)}$ (n = 1, 2, 3). For Type B, P_1 and P_2 are the properties of Layer 3 and Layer 1, respectively. The discussion of the two types of FGM sandwich plates, Type A with a hard core and Type B with a soft core, will be presented in subsequent sections. In this study, the Poisson's ratio of the plate is considered constant, as its impact on deformation is deemed significantly less impactful than that of Young's modulus [41].

2.3. Displacement Field and Strains

In the multi-directional FGM sandwich plate, the in-plane displacements u and v are accounted for in the x and y directions, respectively, while the transverse displacement w occurs in the z direction. These displacements can be expressed as follows, using refined shear deformation theory, as shown in Equations (3)–(6):

$$\begin{aligned} u(x,y,z,t) &= u_0(x,y,t) - z\frac{\partial w_b}{\partial x} - f(z)\frac{\partial w_b}{\partial x} \\ v(x,y,z,t) &= v_0(x,y,t) - z\frac{\partial w_b}{\partial y} - f(z)\frac{\partial w_b}{\partial y} \\ w(x,y,z,t) &= w_b(x,y,t) + w_s(x,y,t) \end{aligned} \tag{3}$$

In this study, two different shape functions are considered, as follows:

$$\begin{aligned} f(z) &= -\tfrac{1}{4}z + \tfrac{5}{3}z\left(\tfrac{z}{h}\right)^2 \text{ for Model 1} \\ f(z) &= z - \tfrac{h}{\pi}\sin\left(\tfrac{\pi z}{h}\right) \text{ for Model 2} \end{aligned} \tag{4}$$

where u_0, v_0, w_b, and w_s are the in-plane and transverse displacements of the middle plane. The strains associated with the displacements in Equation (3) are given by Equations (5) and (6):

$$\begin{aligned} \varepsilon_x &= \varepsilon_x^0 + z\,k_x^b + f\,k_x^s \\ \varepsilon_y &= \varepsilon_y^0 + z\,k_y^b + f\,k_y^s \\ \gamma_{xy} &= \gamma_{xy}^0 + z\,k_{xy}^b + f\,k_{xy}^s \\ \gamma_{yz} &= g\,\gamma_{yz}^s \\ \gamma_{xz} &= g\,\gamma_{xz}^s \\ \varepsilon_z &= 0 \end{aligned} \tag{5}$$

where

$$\begin{aligned} &\varepsilon_x^0 = \frac{\partial u_0}{\partial x},\ k_x^b = -\frac{\partial^2 w_b}{\partial x^2},\ k_x^s = -\frac{\partial^2 w_s}{\partial x^2} \\ &\varepsilon_y^0 = \frac{\partial v_0}{\partial y},\ k_y^b = -\frac{\partial^2 w_b}{\partial y^2},\ k_y^s = -\frac{\partial^2 w_s}{\partial y^2} \\ &\gamma_{xy}^0 = \frac{\partial u_0}{\partial y} + \frac{\partial v_0}{\partial x},\ k_{xy}^b = -2\frac{\partial^2 w_b}{\partial x \partial y},\ k_{xy}^s = -2\frac{\partial^2 w_s}{\partial x \partial y} \\ &\gamma_{yz}^s = \frac{\partial w_s}{\partial y},\ \gamma_{xz}^s = \frac{\partial w_s}{\partial x} \\ &g(z) = 1 - f'(z),\ f'(z) = \frac{df(z)}{dz} \end{aligned} \tag{6}$$

The stress–strain relationship of a multi-directional FGM sandwich plate can be expressed as shown in Equations (7) and (8):

$$\begin{aligned} &\begin{Bmatrix} \sigma_x \\ \sigma_y \\ \tau_{xy} \end{Bmatrix}^{(n)} = \begin{bmatrix} Q_{11} & Q_{12} & 0 \\ Q_{12} & Q_{22} & 0 \\ 0 & 0 & Q_{66} \end{bmatrix}^{(n)} \begin{Bmatrix} \varepsilon_x \\ \varepsilon_y \\ \gamma_{xy} \end{Bmatrix} \\ \text{and } &\begin{Bmatrix} \tau_{yz} \\ \tau_{zx} \end{Bmatrix}^{(n)} = \begin{bmatrix} Q_{44} & 0 \\ 0 & Q_{55} \end{bmatrix}^{(n)} \begin{Bmatrix} \gamma_{yz} \\ \gamma_{zx} \end{Bmatrix} \end{aligned} \tag{7}$$

where

$$\begin{aligned} &Q_{11}^{(n)}(x,z) = Q_{22}^{(n)}(x,z) = \frac{E^{(n)}(x,z)}{1-\nu^2} \\ &Q_{12} = \nu \cdot Q_{11}^{(n)}(x,z) \\ &Q_{44}^{(n)}(x,z) = Q_{55}^{(n)} = Q_{66}^{(n)} = \frac{E^{(n)}(x,z)}{2(1+\nu)} \end{aligned} \tag{8}$$

2.4. Governing Equations

Hamilton's principle (Equation (9)) is employed here to derive the equations of motion:

$$0 = \int_0^t (\delta U + \delta V - \delta K)\, dt \tag{9}$$

Here, δU represents the variation of strain energy, δV denotes the variation of work done, and δK signifies the variation of kinetic energy. The variation of the strain energy of the plate is expressed as shown in Equation (10):

$$\begin{aligned}\delta U &= \int_V \left[\sigma_x \delta\, \varepsilon_x + \sigma_y \delta\, \varepsilon_y + \tau_{xy} \delta\, \gamma_{xy} + \tau_{yz} \delta\, \gamma_{yz} + \tau_{xz} \delta\, \gamma_{xz}\right] dV \\ &= \int_A \Big[N_x \delta\, \varepsilon_x^0 + N_y \delta\, \varepsilon_y^0 + N_{xy} \delta\, \gamma_{xy}^0 + M_x^b \delta\, k_x^b + M_y^b \delta\, k_y^b + M_{xy}^b \delta\, k_{xy}^b \\ &+ M_x^s \delta\, k_x^s + M_y^s \delta\, k_y^s + M_{xy}^s \delta\, k_{xy}^s + S_{yz}^s \delta\, \gamma_{yz}^s + S_{xz}^s \delta\, \gamma_{xz}^s\Big]\, dA = 0\end{aligned} \tag{10}$$

where A is the top surface and the stress resultants N, M, and S are defined as

$$\begin{aligned}\left(N_i, M_i^b, M_i^s\right) &= \sum_{n=1}^{3} \int_{h_n}^{h_{n+1}} (1, z, f)\sigma_i dz,\ (i = x, y, xy) \\ \left(S_{xz}^s, S_{yz}^s\right) &= \sum_{n=1}^{3} \int_{h_n}^{h_{n+1}} g(\tau_{xz}, \tau_{yz}) dz\end{aligned} \tag{11}$$

The variation of the work done by the in-plane load $\left(N_x^0, N_y^0, N_{xy}^0\right)$ can be expressed as

$$\delta V = -\int_A \overline{N} \delta(w_b + w_s) dA \tag{12}$$

with

$$\overline{N} = \left[N_x^0 \frac{\partial^2 (w_b + w_s)}{\partial x^2} + N_y^0 \frac{\partial^2 (w_b + w_s)}{\partial y^2} + 2N_{xy}^0 \frac{\partial^2 (w_b + w_s)}{\partial x \partial y}\right] \tag{13}$$

The variation in the kinetic energy of the plate can be expressed as

$$\begin{aligned}\delta K = &\int_{-h/2}^{h/2} \int_A (\dot{u} \delta\, \dot{u} + \dot{v} \delta\, \dot{v} + \dot{w} \delta\, \dot{w})\, \rho(z)\, dA\, dz \\ = &\int_A \Big\{ I_0 \left[\dot{u}_0 \delta \dot{u}_0 + \dot{v}_0 \delta \dot{v}_0 + (\dot{w}_b + \dot{w}_s)(\delta \dot{w}_b + \delta \dot{w}_s)\right] \\ &- I_1 \left(\dot{u}_0 \tfrac{\partial \delta \dot{w}_b}{\partial x} + \tfrac{\partial \dot{w}_b}{\partial x} \delta\, \dot{u}_0 + \dot{v}_0 \tfrac{\partial \delta \dot{w}_b}{\partial y} + \tfrac{\partial \dot{w}_b}{\partial y} \delta\, \dot{v}_0\right) \\ &- I_2 \left(\dot{u}_0 \tfrac{\partial \delta \dot{w}_s}{\partial x} + \tfrac{\partial \dot{w}_s}{\partial x} \delta\, \dot{u}_0 + \dot{v}_0 \tfrac{\partial \delta \dot{w}_s}{\partial y} + \tfrac{\partial \dot{w}_s}{\partial y} \delta\, \dot{v}_0\right) \\ &+ J_1 \left(\tfrac{\partial \dot{w}_b}{\partial x} \tfrac{\partial \delta\, \dot{w}_b}{\partial x} + \tfrac{\partial \dot{w}_b}{\partial y} \tfrac{\partial \delta\, \dot{w}_b}{\partial y}\right) + K_2 \left(\tfrac{\partial \dot{w}_s}{\partial x} \tfrac{\partial \delta\, \dot{w}_s}{\partial x} + \tfrac{\partial \dot{w}_s}{\partial y} \tfrac{\partial \delta\, \dot{w}_s}{\partial y}\right) \\ &+ J_2 \left(\tfrac{\partial \dot{w}_b}{\partial x} \tfrac{\partial \delta\, \dot{w}_s}{\partial x} + \tfrac{\partial \dot{w}_s}{\partial x} \tfrac{\partial \delta\, \dot{w}_b}{\partial x} + \tfrac{\partial \dot{w}_b}{\partial y} \tfrac{\partial \delta\, \dot{w}_s}{\partial y} + \tfrac{\partial \dot{w}_s}{\partial y} \tfrac{\partial \delta\, \dot{w}_b}{\partial y}\right) \Big\} dA\end{aligned} \tag{14}$$

The notation with a dot superscript denotes differentiation with respect to the time variable t, where $\rho(z)$ represents the mass density defined by Equation (3) and (I_i, J_i, K_i) denote mass inertias, expressed as

$$\begin{aligned}(I_0, I_1, I_2) &= \sum_{n=1}^{3} \int_{h_n}^{h_{n+1}} (1, z, z^2)\rho(z) dz \\ (J_1, J_2, K_2) &= \sum_{n=1}^{3} \int_{h_n}^{h_{n+1}} (f, z\, f, f^2)\rho(z) dz\end{aligned} \tag{15}$$

By substituting Equations (10), (12), and (14) into Equation (9), the following can be derived:

$$\begin{aligned}
&\delta u_0: \frac{\partial N_x}{\partial x}+\frac{\partial N_{xy}}{\partial y}=I_0\ddot{u}_0-I_1\frac{\partial \ddot{w}_b}{\partial x}-J_1\frac{\partial \ddot{w}_s}{\partial x}\\
&\delta v_0: \frac{\partial N_{xy}}{\partial x}+\frac{\partial N_y}{\partial y}=I_0\ddot{v}_0-I_1\frac{\partial \ddot{w}_b}{\partial y}-J_1\frac{\partial \ddot{w}_s}{\partial y}\\
&\delta w_b: \frac{\partial^2 M_x^b}{\partial x^2}+2\frac{\partial^2 M_{xy}^b}{\partial x\partial y}+\frac{\partial^2 M_y^b}{\partial y^2}+\overline{N}=I_0(\ddot{w}_b+\ddot{w}_s)+I_1\left(\frac{\partial \ddot{u}_0}{\partial x}+\frac{\partial \ddot{v}_0}{\partial y}\right)-I_2\nabla^2\ddot{w}_b-J_2\nabla^2\ddot{w}_s\\
&\delta w_s: \frac{\partial^2 M_x^s}{\partial x^2}+2\frac{\partial^2 M_{xy}^s}{\partial x\partial y}+\frac{\partial^2 M_y^s}{\partial y^2}+\frac{\partial S_{xz}^s}{\partial x}+\frac{\partial S_{yz}^s}{\partial y}+\overline{N}=I_0(\ddot{w}_b+\ddot{w}_s)+J_1\left(\frac{\partial \ddot{u}_0}{\partial x}+\frac{\partial \ddot{v}_0}{\partial y}\right)-J_2\nabla^2\ddot{w}_b-K_2\nabla^2\ddot{w}_s
\end{aligned} \tag{16}$$

By substituting Equation (7) into Equation (11) and integrating across the thickness of the plate, the stress resultants can be expressed compactly in terms of strains as follows:

$$\begin{Bmatrix} N \\ M^b \\ M^s \end{Bmatrix}=\begin{bmatrix} A & B & B^s \\ B & D & D^s \\ B^s & D^s & H^s \end{bmatrix}\begin{Bmatrix} \varepsilon \\ k^b \\ k^s \end{Bmatrix} \tag{17}$$

in which

$$\begin{aligned}
&N=\{N_x,N_y,N_{xy}\}^t,\ M^b=\left\{M_x^b,M_y^b,M_{xy}^b\right\}^t,\ M^s=\left\{M_x^s,M_y^s,M_{xy}^s\right\}^t\\
&\varepsilon=\left\{\varepsilon_x^0,\varepsilon_y^0,\gamma_{xy}^0\right\}^t,\ k^b=\left\{k_x^b,k_y^b,k_{xy}^b\right\}^t,\ k^s=\left\{k_x^s,k_y^s,k_{xy}^s\right\}^t\\
&A=\begin{bmatrix} A_{11} & A_{12} & 0 \\ A_{12} & A_{22} & 0 \\ 0 & 0 & A_{66} \end{bmatrix},\ B=\begin{bmatrix} B_{11} & B_{12} & 0 \\ B_{12} & B_{22} & 0 \\ 0 & 0 & B_{66} \end{bmatrix},\ D=\begin{bmatrix} D_{11} & D_{12} & 0 \\ D_{12} & D_{22} & 0 \\ 0 & 0 & D_{66} \end{bmatrix}\\
&B^s=\begin{bmatrix} B_{11}^s & B_{12}^s & 0 \\ B_{12}^s & B_{22}^s & 0 \\ 0 & 0 & B_{66}^s \end{bmatrix},\ D^s=\begin{bmatrix} D_{11}^s & D_{12}^s & 0 \\ D_{12}^s & D_{22}^s & 0 \\ 0 & 0 & D_{66}^s \end{bmatrix},\ H^s=\begin{bmatrix} H_{11}^s & H_{12}^s & 0 \\ H_{12}^s & H_{22}^s & 0 \\ 0 & 0 & H_{66}^s \end{bmatrix}\\
&S=\left\{S_{xz}^s,S_{yz}^s\right\}^t,\ \gamma=\left\{\gamma_{xz}^0,\gamma_{yz}^0\right\}^t,\ A^s=\begin{bmatrix} A_{44}^s & 0 \\ 0 & A_{55}^s \end{bmatrix}
\end{aligned} \tag{18}$$

The stiffness components are given as

$$\begin{aligned}
&\begin{Bmatrix} A_{11} & B_{11} & D_{11} & B_{11}^s & D_{11}^s & H_{11}^s \\ A_{12} & B_{12} & D_{12} & B_{12}^s & D_{12}^s & H_{12}^s \\ A_{66} & B_{66} & D_{66} & B_{66}^s & D_{66}^s & H_{66}^s \end{Bmatrix}=\int_{-h/2}^{h/2} Q_{11}\left(1,z,z^2,f(z),z\,f(z),f^2(z)\right)\begin{Bmatrix} 1 \\ \nu \\ \frac{1-\nu}{2} \end{Bmatrix}dz\\
&\left(A_{22},B_{22},D_{22},B_{22}^s,D_{22}^s,H_{22}^s\right)=\left(A_{11},B_{11},D_{11},B_{11}^s,D_{11}^s,H_{11}^s\right)\\
&A_{44}^s=A_{55}^s=\int_{-h/2}^{h/2} Q_{44}[g(z)]^2dz
\end{aligned} \tag{19}$$

By introducing Equation (17) into Equation (16), the equations of motion can be rephrased in terms of displacements (u_0, v_0, w_b, w_s). The resulting equations take the following forms:

$$\begin{aligned}
&A_{11}\frac{\partial^2 u_0}{\partial x^2}+A_{66}\frac{\partial^2 u_0}{\partial y^2}+(A_{12}+A_{66})\frac{\partial^2 v}{\partial x\partial y}-B_{11}\frac{\partial^3 w_b}{\partial x^3}-(B_{12}+2B_{66})\frac{\partial^3 w_b}{\partial x\partial y^2}\\
&-B_{11}^s\frac{\partial^3 w_s}{\partial x^3}-\left(B_{12}^s+2B_{66}^s\right)\frac{\partial^3 w_s}{\partial x\partial y^2}=I_0\ddot{u}_0-I_1\frac{\partial \ddot{w}_b}{\partial x}-J_1\frac{\partial \ddot{w}_s}{\partial x}\\
&(A_{12}+A_{66})\frac{\partial^2 u_0}{\partial x\partial y}+A_{66}\frac{\partial^2 v_0}{\partial x^2}+A_{22}\frac{\partial^2 v_0}{\partial y^2}-(B_{12}+2B_{66})\frac{\partial^3 w_b}{\partial x^2\partial y}-B_{22}\frac{\partial^3 w_b}{\partial y^3}\\
&-B_{22}^s\frac{\partial^3 w_s}{\partial y^3}-\left(B_{12}^s+2B_{66}^s\right)\frac{\partial^3 w_s}{\partial x^2\partial y}=I_0\ddot{v}_0-I_1\frac{\partial \ddot{w}_b}{\partial y}-J_1\frac{\partial \ddot{w}_s}{\partial y}\\
&B_{11}\frac{\partial^3 u_0}{\partial x^3}+(B_{12}+2B_{66})\frac{\partial^3 u_0}{\partial x\partial y^2}+(B_{12}+2B_{66})\frac{\partial^3 v_0}{\partial x^2\partial y}+B_{22}\frac{\partial^3 v_0}{\partial y^3}-D_{11}\frac{\partial^4 w_b}{\partial x^4}\\
&-2(D_{12}+2D_{66})\frac{\partial^4 w_b}{\partial x^2\partial y^2}-D_{22}\frac{\partial^4 w_b}{\partial y^4}-D_{11}^s\frac{\partial^4 w_s}{\partial x^4}-2\left(D_{12}^s+2D_{66}^s\right)\frac{\partial^4 w_s}{\partial x^2\partial y^2}\\
&-D_{22}^s\frac{\partial^4 w_s}{\partial y^4}+\overline{N}=I_0(\ddot{w}_b+\ddot{w}_s)+I_1\left(\frac{\partial \ddot{u}_0}{\partial x}+\frac{\partial \ddot{v}_0}{\partial y}\right)-I_2\nabla^2\ddot{w}_b-J_2\nabla^2\ddot{w}_s\\
&B_{11}^s\frac{\partial^3 u_0}{\partial x^3}+\left(B_{12}^s+2B_{66}^s\right)\frac{\partial^3 u_0}{\partial x\partial y^2}+\left(B_{12}^s+2B_{66}^s\right)\frac{\partial^3 v_0}{\partial x^2\partial y}+B_{22}^s\frac{\partial^3 v_0}{\partial y^3}-D_{11}^s\frac{\partial^4 w_b}{\partial x^4}\\
&-2\left(D_{12}^s+2D_{66}^s\right)\frac{\partial^4 w_b}{\partial x^2\partial y^2}-D_{22}^s\frac{\partial^4 w_b}{\partial y^4}-H_{11}^s\frac{\partial^4 w_s}{\partial x^4}-2\left(H_{12}^s+2H_{66}^s\right)\frac{\partial^4 w_s}{\partial x^2\partial y^2}-H_{22}^s\frac{\partial^4 w_s}{\partial y^4}\\
&+A_{55}^s\frac{\partial^2 w_s}{\partial x^2}+A_{44}^s\frac{\partial^2 w_s}{\partial y^2}+\overline{N}=I_0(\ddot{w}_b+\ddot{w}_s)+J_1\left(\frac{\partial \ddot{u}_0}{\partial x}+\frac{\partial \ddot{v}_0}{\partial y}\right)-J_2\nabla^2\ddot{w}_b-K_2\nabla^2\ddot{w}_s
\end{aligned} \tag{20}$$

The exact solution to the set of Equation (20), describing the P-FGMs multi-directional functionality graded sandwich plate under various boundary conditions, can be derived. The boundary conditions for an arbitrary edge include both its simply supported and clamped conditions:

- Clamped (C):

$$u_0 = v_0 = w_b = \partial w_b/\partial x = \partial w_b/\partial y = w_s = \partial w_s/\partial x = \partial w_s/\partial y = 0, \quad \text{at } x = 0, y = 0, b \tag{21}$$

- Simply supported (S):

$$\begin{aligned} v_0 = w_b = \partial w_b/\partial y = w_s = \partial w_s/\partial y = 0, \quad \text{at } x = 0, a \\ u_0 = w_b = \partial w_b/\partial x = w_s = \partial w_s/\partial x = 0, \quad \text{at } y = 0, b \end{aligned} \tag{22}$$

The following representation of the displacement quantities that fulfill the aforementioned boundary conditions is applicable to our specific problem:

$$\begin{Bmatrix} u_0 \\ v_0 \\ w_b \\ w_s \end{Bmatrix} = \begin{Bmatrix} U_{mn}\frac{\partial X_m(x)}{\partial x}Y_n(y)e^{i\omega t} \\ V_{mn}X_n(x)\frac{\partial Y_n(y)}{\partial y}e^{i\omega t} \\ W_{bmn}X_n(x)Y_n(y)e^{i\omega t} \\ W_{smn}X_n(x)Y_n(y)e^{i\omega t} \end{Bmatrix} \tag{23}$$

where U_{mn}, V_{mn}, W_{bmn}, and W_{smn} are arbitrary parameters and $\omega = \omega_{mn}$ denotes the eigenfrequency associated with the $(m,n)^{th}$ eigenmode. The functions $X_m(x)$ and $Y_n(y)$ are suggested here to satisfy, at least, the geometric boundary conditions given in Equations (21) and (22) and represent the approximate shapes of the deflected surface of the plate. These functions, for the different boundary condition cases, are listed in Table 1. Note that $\lambda = m\pi/a$ and $\mu = n\pi/b$.

Table 1. The admissible functions for the various boundary conditions [42].

Boundary Conditions	$x = 0$	$y = 0$	$x = a$	$y = b$	$X_m(x)$	$Y_n(y)$
SSSS	S	S	S	S	$\sin(\lambda x)$	$\sin(\mu x)$
CSCS	C	S	C	S	$\sin^2(\lambda x)$	$\sin(\mu x)$
CCCC	C	C	C	C	$\sin^2(\lambda x)$	$\sin^2(\mu x)$
FCFC	F	C	F	C	$\cos^2(\lambda x)\cdot[\sin^2(\lambda x) + 1]$	$\sin^2(\mu x)$

By substituting Equation (23) into the governing Equation (20) and multiplying each equation by its corresponding eigenfunction, and then integrating over the solution domain, we can derive the following equations after performing certain mathematical manipulations:

$$\left(\begin{bmatrix} a_{11} & a_{12} & a_{13} & a_{14} \\ a_{21} & a_{22} & a_{23} & a_{24} \\ a_{31} & a_{32} & a_{33} - \beta N & a_{34} - \beta N \\ a_{41} & a_{42} & a_{43} - \beta N & a_{44} - \beta N \end{bmatrix} - \omega^2 \begin{bmatrix} m_{11} & 0 & m_{13} & m_{14} \\ 0 & m_{22} & m_{23} & m_{24} \\ m_{31} & m_{32} & m_{33} & m_{34} \\ m_{41} & m_{42} & m_{43} & m_{44} \end{bmatrix} \right) \cdot \begin{Bmatrix} U_{mn} \\ V_{mn} \\ W_{mn} \\ X_{mn} \end{Bmatrix} = \begin{Bmatrix} 0 \\ 0 \\ 0 \\ 0 \end{Bmatrix} \tag{24}$$

in which

$$
\begin{array}{llll}
a_{11}=A_{11}\alpha_{12}+A_{66}\alpha_8 & a_{12}=(A_{12}+A_{66})\alpha_8 & a_{13}=-B_{11}\alpha_{12}-(B_{12}+2B_{66})\alpha_8 & a_{14}=-(B^s_{12}+2B^s_{66})\alpha_8-B^s_{11}\alpha_{12} \\
a_{21}=(A_{12}+A_{66})\alpha_{10} & a_{22}=A_{22}\alpha_4+A_{66}\alpha_{10} & a_{23}=-B_{22}\alpha_4-(B_{12}+2B_{66})\alpha_{10} & a_{24}=-(B^s_{12}+2B_{66})\alpha_{10}-B^s_{22}\alpha_4 \\
a_{31}=B_{11}\alpha_{13}+(B_{12}+2B_{66})\alpha_{11} & a_{32}=(B_{12}+2B_{66})\alpha_{11} & a_{33}=-D_{11}\alpha_{13}-2(D_{12}+2D_{66})\alpha_{11} & a_{34}=-D^s_{11}\alpha_{13}-2(D^s_{12}+2D^s_{66})\alpha_{11} \\
 & +B_{22}\alpha_5 & -D_{22}\alpha_5 & -D^s_{66}\alpha_5 \\
a_{41}=B^s_{11}\alpha_{13}+(B^s_{12}+2B^s_{66})\alpha_{11} & a_{42}=(B^s_{12}+2B^s_{66})\alpha_{11} & a_{43}=-D^s_{11}\alpha_{13}-2(D^s_{12}+2D^s_{66})\alpha_{11} & a_{44}=-H^s_{11}\alpha_{13}-2(H^s_{12}+2H^s_{66})\alpha_{11} \\
 & +B^s_{22}\alpha_5 & -D^s_{22}\alpha_5 & -H^s_{22}\alpha_5+A^s_{44}\alpha_9+A^s_{55}\alpha_3 \\
N=N^0_x & & & \\
\xi_1=N^0_y/N^0_x & & &
\end{array}
\tag{25}
$$

and

$$
\begin{aligned}
m_{11} &= -I_0\alpha_6 \\
m_{13} &= -I_1\alpha_6,\ m_{32} = -I_1\alpha_3 \\
m_{14} &= J_1\alpha_6,\ m_{33} = -I_0\alpha_1 + I_2(\alpha_3+\alpha_9) \\
m_{22} &= -I_0\alpha_2,\ m_{34} = -I_0\alpha_1 + J_2(\alpha_3+\alpha_9) \\
m_{23} &= I_1\alpha_2,\ m_{41} = -J_1\alpha_9 \\
m_{24} &= J_1\alpha_2,\ m_{42} = -J_1\alpha_3 \\
m_{31} &= -I_1\alpha_9,\ m_{44} = -I_0\alpha_1 + K_2(\alpha_3+\alpha_9)
\end{aligned}
\tag{26}
$$

with

$$
\begin{aligned}
&\beta = \xi_1\alpha_3 + \alpha_9 \\
&(\alpha_1,\alpha_3,\alpha_5) = \int_0^b\int_0^a (X_mY_n, X_mY''_n, X_mY'''_n)X_mY_n dxdy \\
&(\alpha_2,\alpha_4,\alpha_{10}) = \int_0^b\int_0^a (X_mY'_n, X_mY''_n, X''_mY'_n)X_mY'_n dxdy \\
&(\alpha_6,\alpha_8,\alpha_{12}) = \int_0^b\int_0^a (X'_mY_n, X'_mY''_n, X'''_mY_n)X'_mY_n dxdy \\
&(\alpha_7,\alpha_9,\alpha_{11},\alpha_{13}) = \int_0^b\int_0^a (X'_mY'_n, X''_mY_n, X'_mY''_n, X'''_mY_n)X_mY_n dxdy
\end{aligned}
\tag{27}
$$

3. Numerical Results and Discussion

In this section, we explore several numerical examples to assess the accuracy of the two proposed theories in analyzing the buckling and free vibration of multi-directional FG sandwich plates under different boundary conditions. A range of sandwich plate configurations, comprising both symmetric and non-symmetric FGMs, are examined to demonstrate the versatility of these theories.

The subsequent discussions aim to underscore the precision and relevance of the presented theories in capturing the plate's behavior under varied conditions. The considered configuration schemes of the sandwich plates include:

1. **(1-0-1)** FGM sandwich plate, consisting of two layers of equal thickness without a core, where $h_1 = h_2 = 0$.
2. **(1-2-1)** FGM sandwich plate, with the core thickness equal to the sum of the face thicknesses: $h_2 = -h/4, h_3 = h/4$.
3. **(1-1-1)** FGM sandwich plate, comprising three equal-thickness layers: $h_2 = -h/6$, $h_3 = h/6$.
4. **(2-2-1)** FGM sandwich plate, featuring a core thickness twice that of the upper face and equal to the lower one, defined by $h_2 = -h/10, h_3 = 3h/10$.
5. **(1-1-2)** FGM sandwich plate, with a core thickness equal to that of the lower face and with the thickness of the upper face twice that of the core: $h_2 = -h/4, h_3 = 0$.

The material combinations include aluminum and alumina, each with the following properties [23]:

- **Ceramic** (alumina, Al_2O_3): Young's modulus $E_c = 380$ GPa, Poisson's ratio $v_c = 0.3$, density $\rho_c = 3800\ \mathrm{kg/m^3}$;

- **Metal** (aluminum, Al): Young's modulus E_m = 70 GPa, Poisson's ratio υ_m = 0.3, density ρ_m = 2702 kg/m^3.

3.1. FG Sandwich Plates

3.1.1. Free Vibration Analysis of FG Sandwich Plates

The face sheet comprises a functionally graded material with properties varying along the p_z direction. A power law distribution is utilized for the FG face sheet, while the core consists of a homogeneous material. When the core material is pure ceramic (alumina), it is referred to as "hard-core", and when it is pure metal (aluminum) it is called "soft-core". Table 2 displays the fundamental frequency results obtained from the two proposed methods for a/h = 10 and p_x = 0, in the case of a hard core. The results indicate that the sandwich plate with the (1-2-1) scheme exhibits the highest frequency because of its thicker ceramic core, leading to greater stiffness. Subsequently, a decrease in natural frequency was observed for the (2-2-1) scheme, followed by the (1-1-1), (2-1-2), and (1-0-1) schemes. Their frequency diminishes with an increase in the grading parameter p_z.

Table 2. Dimensionless fundamental frequency $\overline{\omega}$ of FG sandwich plates (a/h = 10 and p_x = 0). Type A: hard core.

p_z	Theory	1-0-1	2-1-2	1-1-1	2-2-1	1-2-1
0	3D [43]	1.8268	1.8268	1.8268	1.8268	1.8268
	SSDT [44]	1.8245	1.8245	1.8245	1.8245	1.8245
	TSDT [44]	1.8245	1.8245	1.8245	1.8245	1.8245
	FSDT [44]	1.8244	1.8244	1.8244	1.8244	1.8244
	NFSDT [45]	1.8244	1.8244	1.8244	1.8244	1.8244
	Present 1	**1.8245**	**1.8245**	**1.8245**	**1.8245**	**1.8245**
	Present 2	**1.8245**	**1.8245**	**1.8245**	**1.8245**	**1.8245**
0.5	3D [43]	1.4461	1.4861	1.5213	1.5493	1.5767
	SSDT [44]	1.4444	1.4842	1.5193	1.5520	1.5745
	TSDT [44]	1.4442	1.4841	1.5192	1.5520	1.5727
	FSDT [44]	1.4417	1.4816	1.5170	1.5500	1.5727
	NFSDT [45]	1.4442	1.4841	1.5192	1.5471	1.5745
	Present 1	**1.4446**	**1.4844**	**1.5195**	**1.5474**	**1.5747**
	Present 2	**1.4447**	**1.4845**	**1.5195**	**1.5474**	**1.5747**
1	3D [43]	1.2447	1.3018	1.3552	1.3976	1.4414
	SSDT [44]	1.2434	1.3002	1.3534	1.4079	1.4393
	TSDT [44]	1.2432	1.3001	1.3533	1.4079	1.4393
	FSDT [44]	1.2403	1.2973	1.3507	1.4056	1.4372
	NFSDT [45]	1.2429	1.3000	1.3533	1.3956	1.4393
	Present 1	**1.2437**	**1.3005**	**1.3537**	**1.3959**	**1.4396**
	Present 2	**1.2438**	**1.3006**	**1.3537**	**1.3959**	**1.4396**
5	3D [43]	0.9448	0.9810	1.0453	1.1098	1.1757
	SSDT [44]	0.9463	0.9821	1.0448	1.1474	1.1740
	TSDT [44]	0.9460	0.9818	1.0447	1.1473	1.1740
	FSDT [44]	0.9426	0.9787	1.0418	1.1447	1.1716
	NFSDT [45]	0.9431	0.9796	1.0435	1.1077	1.1735
	Present 1	**0.9467**	**0.9824**	**1.0451**	**1.1094**	**1.1743**
	Present 2	**0.9469**	**0.9826**	**1.0453**	**1.1095**	**1.1744**

Additionally, an analysis was conducted for a/h = 5 and p_x = 0 for both hard-core and soft-core materials, as presented in Tables 3 and 4. The analysis reveals that the soft-core material yields lower frequencies than the hard-core material for homogeneous materials. Moreover, as the plate thickness transitions from a/h = 10 to a/h = 5, there is a noticeable decrease in the frequency outcomes.

Table 3. Dimensionless fundamental frequency $\overline{\omega}$ of FG sandwich plates (a/h = 5 and n_x = 0). Type A: hard core.

n_z	Theory	1-0-1	2-1-2	1-1-1	2-2-1	1-2-1
0	3D [43]	1.6771	1.6771	1.6771	1.6771	1.6771
	NFSDT [45]	1.6697	1.6697	1.6697	1.6697	1.6697
	Present 1	**1.6701**	**1.6701**	**1.6701**	**1.6701**	**1.6701**
0.5	3D [43]	1.3536	1.3905	1.4218	1.4454	1.4694
	NFSDT [45]	1.3473	1.3841	1.4152	1.4386	1.4626
	Present 1	**1.3478**	**1.3844**	**1.4154**	**1.4388**	**1.4628**
1	3D [43]	1.1749	1.2292	1.2777	1.3143	1.3534
	NFSDT [45]	1.1691	1.2232	1.2714	1.3078	1.3467
	Present 1	**1.1703**	**1.2238**	**1.2717**	**1.3082**	**1.3471**
5	3D [43]	0.8909	0.9336	0.9980	1.0561	1.1190
	NFSDT [45]	0.8853	0.9286	0.9916	1.0488	1.1118
	Present 1	**0.89528**	**0.9365**	**0.9959**	**1.0533**	**1.1136**
10	3D [43]	0.8683	0.8923	0.9498	1.0095	1.0729
	NFSDT [45]	0.8599	0.8860	0.9428	1.0012	1.0648
	Present 1	**0.8725**	**0.8998**	**0.9508**	**1.0095**	**1.0679**

Table 4. Dimensionless fundamental frequency $\overline{\omega}$ of FG sandwich plates (a/h = 5 and n_x = 0). Type B: soft core.

n_z	Theory	1-0-1	2-1-2	1-1-1	2-2-1	1-2-1
0	3D [43]	0.8529	0.8529	0.8529	0.8529	0.8529
	NFSDT [45]	0.8491	0.8491	0.8491	0.8491	0.8491
	Present 1	**0.8501**	**0.8501**	**0.8501**	**0.8501**	**0.8501**
0.5	3D [43]	1.3789	1.3206	1.2805	1.2453	1.2258
	NFSDT [45]	1.3686	1.3115	1.2729	1.2380	1.2185
	Present 1	**1.3829**	**1.3284**	**1.2859**	**1.2509**	**1.2255**
1	3D [43]	1.5090	1.4333	1.3824	1.3420	1.3213
	NFSDT [45]	1.4915	1.4156	1.3702	1.3302	1.3104
	Present 1	**1.5176**	**1.4557**	**1.4036**	**1.3625**	**1.3289**
5	3D [43]	1.6587	1.5801	1.5028	1.4601	1.4267
	NFSDT [45]	1.6305	1.5125	1.4589	1.4195	1.4026
	Present 1	**1.6585**	**1.6181**	**1.5665**	**1.5212**	**1.4748**
10	3D [43]	1.6728	1.6091	1.5267	1.4831	1.4410
	NFSDT [45]	1.6495	1.5196	1.4642	1.4266	1.4101
	Present 1	**1.6679**	**1.6394**	**1.5931**	**1.5484**	**1.5018**

When n_z = 0.5, the hard-core plate exhibits higher frequencies than the soft-core plate for all instances, except for the 1-0-1 scheme, where the soft core demonstrates higher frequencies. Conversely, at n_z = 1, the soft core demonstrates higher frequencies than the hard core, except for the 1-2-1 scheme. Beyond n_z > 5, it is observed that the soft-core material consistently yields higher frequencies than the hard-core material.

3.1.2. Buckling Analysis of FG Sandwich Plates

The buckling analysis results mirror the trends observed in the frequency analysis, as summarized in Table 5. Similar to the frequency outcomes, the buckling analysis focuses on a square plate with a hard core material undergoing uniaxial compression, particularly at a/h = 10. Among the different schemes examined, the (1-2-1) configuration demonstrates the highest buckling load, followed by (2-2-1), (1-1-1), (2-1-2), and (1-0-1), in descending order of buckling strength. Notably, there is a consistent decrease in buckling load with the increasing grading parameter p_z. This pattern underscores the significant influence of the grading parameter on the buckling behavior of square plates with a hard core material under uniaxial compression, with various schemes exhibiting distinct levels of structural stability.

Table 5. Dimensionless buckling load $\overline{N}$ of square plates under uniaxial compression ($\xi_1 = 0$, $a/h = 10$). Type A: hard core.

n_z	Theory	1-0-1	2-1-2	1-1-1	2-2-1	1-2-1
0	SSDT [44]	13.0061	13.0061	13.0061	13.0061	13.0061
	TSDT [44]	13.0050	13.0050	13.0050	13.0050	13.0050
	FSDT [44]	13.0045	13.0045	13.0045	13.0045	13.0045
	NFSDT [45]	13.0045	13.0045	13.0045	13.0045	13.0045
	Present 1	**13.0049**	**13.0049**	**13.0049**	**13.0049**	**13.0049**
	Present 2	**13.0061**	**13.0061**	**13.0061**	**13.0061**	**13.0061**
0.5	SSDT [44]	7.3657	7.9420	8.4371	8.8104	9.2167
	TSDT [44]	7.3644	7.9408	8.4365	8.8100	9.2168
	FSDT [44]	7.3373	7.9132	8.4103	8.7867	9.1952
	NFSDT [45]	7.3634	7.9403	8.4361	8.8095	9.2162
	Present 1	**7.3644**	**7.9408**	**8.4365**	**8.8099**	**9.2168**
	Present 2	**7.3657**	**7.9419**	**8.4371**	**8.8104**	**9.2167**
1	SSDT [44]	5.1685	5.8412	6.4654	6.9498	7.5063
	TSDT [44]	5.1671	5.8401	6.4647	6.9494	7.5066
	FSDT [44]	5.1424	5.8138	6.4389	6.9257	7.4837
	NFSDT [45]	5.1648	5.8387	6.4641	6.9485	7.5056
	Present 1	**5.1671**	**5.8401**	**6.4647**	**6.9494**	**7.5066**
	Present 2	**5.1685**	**5.8412**	**6.4654**	**6.9498**	**7.5063**
5	SSDT [44]	2.6601	3.0441	3.5806	4.1129	4.7349
	TSDT [44]	2.6582	3.0426	3.5796	4.1121	4.7347
	FSDT [44]	2.6384	3.0225	3.5596	4.0929	4.7148
	NFSDT [45]	2.6415	3.0282	3.5710	4.1024	4.7305
	Present 1	**2.6582**	**3.0426**	**3.5796**	**4.1121**	**4.7347**
	Present 2	**2.6601**	**3.0441**	**3.5806**	**4.1129**	**4.7349**

Moreover, a buckling analysis of square FG plates under biaxial compression was conducted, as indicated in Table 6. This table provides a comprehensive comparison of the dimensionless buckling loads for square plates across various theories and conditions, offering insights into the impact of the grading parameter on the structural stability of the plate. Similar to unidirectional loading, biaxial loading also results in a noticeable decrease in the dimensionless buckling loads for all the schemes considered.

Table 6. Dimensionless buckling load $\overline{N}$ of square plates under biaxial compression ($\xi_1 = 1$, $a/h = 10$). Type A: hard core.

n_z	Theory	1-0-1	2-1-2	1-1-1	2-2-1	1-2-1
0	SSDT [44]	6.5030	6.5030	6.5030	6.5030	6.5030
	TSDT [44]	6.5025	6.5025	6.5025	6.5025	6.5025
	FSDT [44]	6.5022	6.5022	6.5022	6.5022	6.5022
	NFSDT [45]	6.5022	6.5022	6.5022	6.5022	6.5022
	Present 1	**6.5025**	**6.5025**	**6.5025**	**6.5025**	**6.5025**
	Present 2	**6.5030**	**6.5030**	**6.5030**	**6.5030**	**6.5030**
0.5	SSDT [44]	3.6828	3.9710	4.2186	4.4052	4.6084
	TSDT [44]	3.6822	3.9704	4.2182	4.4050	4.6084
	FSDT [44]	3.6687	3.9566	4.2052	4.3934	4.5976
	NFSDT [45]	3.6817	3.9702	4.2181	4.4047	4.6081
	Present 1	**3.6822**	**3.9704**	**4.2182**	**4.4049**	**4.6084**
	Present 2	**3.6828**	**3.9709**	**4.2185**	**4.4052**	**4.6083**

Table 6. Cont.

n_z	Theory	1-0-1	2-1-2	1-1-1	2-2-1	1-2-1
1	SSDT [44]	2.5842	2.9206	3.2327	3.4749	3.7531
	TSDT [44]	2.5836	2.9200	3.2324	3.4747	3.7533
	FSDT [44]	2.5712	2.9069	3.2195	3.4629	3.7418
	NFSDT [45]	2.5824	2.9193	3.2320	3.4742	3.7528
	Present 1	**2.5835**	**2.9200**	**3.2323**	**3.4747**	**3.7533**
	Present 2	**2.5842**	**2.9206**	**3.2327**	**3.4749**	**3.7531**
5	SSDT [44]	1.3300	1.5220	1.7903	2.0564	2.3674
	TSDT [44]	1.3291	1.5213	1.7898	2.0561	2.3673
	FSDT [44]	1.3192	1.5113	1.7798	2.0464	2.3574
	NFSDT [45]	1.3208	1.5141	1.7855	2.0512	2.3652
	Present 1	**1.3291**	**1.5213**	**1.7898**	**2.0560**	**2.3673**
	Present 2	**1.3300**	**1.5220**	**1.7903**	**2.0564**	**2.3674**

3.2. Multi-Directional FG Sandwich Plates

3.2.1. Free Vibration Analysis of Multi-Directional FG Sandwich Plates

The free vibration analysis of a multi-directional FG skin with a homogeneous hard core was conducted under various boundary conditions. Multi-directional gradation was achieved by adjusting the parameters p_x and p_z. The results are presented in Table 7. Across all boundary conditions, the (1-2-1) scheme exhibits the highest frequency, followed by (2-2-1), (1-1-1), (2-1-2), and (1-0-1). In all scenarios, maintaining a constant and zero value for the parameter p_x while increasing p_z from 0.5 to 1 and 5 results in a decrease in frequency. Similarly, increasing p_x from 0 to 0.5 and up to 2 and then varying p_z also leads to a decreasing trend in frequency.

Table 7. Dimensionless fundamental frequency $\overline{\omega}$ of square plates under various boundary conditions ($a/h = 10$). Type A: hard core.

Boundary Conditions	p_x	p_z	1-0-1	2-1-2	1-1-1	2-2-1	1-2-1
SSSS	0	0.5	1.4446	1.4844	1.5195	1.5474	1.5747
	0	1	1.2437	1.3005	1.3537	1.3959	1.4396
	0	5	0.9467	0.9824	1.0451	1.1094	1.1743
	0.5	0.5	1.3965	1.4302	1.4635	1.4949	1.5216
	0.5	1	1.2113	1.2606	1.3109	1.3556	1.3983
	0.5	5	0.9451	0.9748	1.0338	1.0981	1.1613
	2	0.5	1.2835	1.3046	1.3353	1.3749	1.4024
	2	1	1.1378	1.1708	1.2156	1.2655	1.3076
	2	5	0.9414	0.9590	1.0103	1.0745	1.1341
CSCS	0	0.5	2.1285	2.1868	2.2375	2.2772	2.3165
	0	1	1.8379	1.9218	1.9992	2.0599	2.1231
	0	5	1.4017	1.4583	1.5511	1.6446	1.7399
	0.5	0.5	2.0571	2.1080	2.1569	2.2018	2.2407
	0.5	1	1.7894	1.8634	1.9373	2.0015	2.0638
	0.5	5	1.3983	1.4469	1.5345	1.6280	1.7208
	2	0.5	1.8894	1.9256	1.9718	2.0288	2.0697
	2	1	1.6788	1.7321	1.7988	1.8708	1.9329
	2	5	1.3899	1.4234	1.4999	1.5934	1.6813

Table 7. *Cont.*

Boundary Conditions	p_x	p_z	1-0-1	2-1-2	1-1-1	2-2-1	1-2-1
CCCC	0	0.5	2.6386	2.7106	2.7725	2.8205	2.8684
	0	1	2.2834	2.3877	2.4827	2.5565	2.6338
	0	5	1.7439	1.8180	1.9334	2.0482	2.16583
	0.5	0.5	2.5497	2.6139	2.6744	2.7287	2.7767
	0.5	1	2.2225	2.3157	2.4069	2.4850	2.5616
	0.5	5	1.7387	1.8038	1.9129	2.0277	2.1425
	2	0.5	2.3406	2.3902	2.4486	2.5178	2.5689
	2	1	2.0835	2.1538	2.2371	2.3250	2.4021
	2	5	1.7257	1.7741	1.8701	1.9849	2.0939
FCFC	0	0.5	2.7884	2.8643	2.9290	2.9788	3.0289
	0	1	2.4168	2.5271	2.6268	2.7038	2.7847
	0	5	1.8478	1.9289	2.0510	2.1715	2.2954
	0.5	0.5	2.6941	2.7629	2.8266	2.8830	2.9335
	0.5	1	2.3519	2.4514	2.5474	2.6289	2.7094
	0.5	5	1.8415	1.9137	2.0294	2.1499	2.2709
	2	0.5	2.4724	2.5282	2.5906	2.6628	2.7171
	2	1	2.2035	2.2809	2.3695	2.4614	2.5428
	2	5	1.8257	1.8821	1.9843	2.1048	2.2199

Notably, the results reveal an intriguing observation: an increase in the parameter p_z has a more pronounced effect on the frequency parameter compared to p_x. Thus, enhancing the ceramic composition of the material in its transverse direction demonstrates a more significant influence on frequency than varying the ceramic composition of its longitudinal direction. The frequency is observed to be at its maximum for the FCFC condition, followed by the CCCC, CSCS, and SSSS conditions.

The free vibration analysis of a multi-directional FG skin combined with a homogeneous soft core was also conducted under various boundary conditions for different values of p_x and p_z. The results are presented in Table 8. In contrast to the hard-core sandwich plate, the maximum frequency was observed for the (1-0-1) configuration, followed by (2-1-2), (1-1-1), (2-2-1), and (1-2-1). Similar to the hard-core case, the soft-core frequency is highest for the FCFC condition, followed by the CCCC, CSCS, and SSSS conditions.

Across all boundary conditions, maintaining a fixed and zero value for the parameter p_x while increasing p_z from 0.5 to 1 and 5 results in an increase in frequency. Conversely, increasing p_x from 0 to 0.5 and up to 2 and varying p_z leads to a similar increase in frequency. The results underscore the fact that the parameter p_z has a more significant influence on the frequency parameter compared to p_x.

Table 8. Dimensionless fundamental frequency $\overline{\omega}$ of square plates under various boundary conditions ($a/h = 10$). Type B: soft core.

Boundary Conditions	p_x	p_z	1-0-1	2-1-2	1-1-1	2-2-1	1-2-1
SSSS	0	0.5	1.5758	1.5299	1.4876	1.4372	1.4173
	0	1	1.7263	1.6846	1.6405	1.5798	1.5619
	0	5	1.8422	1.8421	1.8179	1.7541	1.7494
	0.5	0.5	1.6232	1.5879	1.5507	1.4986	1.4828
	0.5	1	1.7452	1.7137	1.6749	1.6146	1.6000
	0.5	5	1.8401	1.8441	1.8228	1.7601	1.7570
	2	0.5	1.7017	1.6887	1.6627	1.6085	1.6006
	2	1	1.7762	1.7652	1.7382	1.6791	1.6714
	2	5	1.8354	1.8476	1.8323	1.7718	1.7720

Table 8. *Cont.*

Boundary Conditions	p_x	p_z	1-0-1	2-1-2	1-1-1	2-2-1	1-2-1
CSCS	0	0.5	2.2702	2.1949	2.1305	2.0639	2.0299
	0	1	2.4889	2.4125	2.3402	2.2608	2.2229
	0	5	2.6799	2.6551	2.6007	2.5159	2.4809
	0.5	0.5	2.3500	2.2855	2.2237	2.1542	2.1209
	0.5	1	2.5258	2.4619	2.3942	2.3144	2.2771
	0.5	5	2.6796	2.6609	2.6105	2.5266	2.4932
	2	0.5	2.4776	2.4404	2.3886	2.3152	2.2845
	2	1	2.5842	2.5476	2.4923	2.4127	2.3788
	2	5	2.6778	2.6715	2.6289	2.5474	2.5171
CCCC	0	0.5	2.7695	2.6701	2.5885	2.5124	2.4665
	0	1	3.0378	2.9312	2.8358	2.7455	2.6897
	0	5	3.2917	3.2401	3.1577	3.0599	2.9948
	0.5	0.5	2.8767	2.7864	2.7043	2.6239	2.5747
	0.5	1	3.0912	2.9977	2.9053	2.8136	2.7552
	0.5	5	3.2937	3.2499	3.1717	3.0748	3.0107
	2	0.5	3.0449	2.9835	2.9084	2.8226	2.7694
	2	1	3.1743	3.1116	3.0305	2.9380	2.8782
	2	5	3.2958	3.2674	3.1982	3.1032	3.0417
FCFC	0	0.5	2.8951	2.7859	2.6987	2.6225	2.5716
	0	1	3.1767	3.0559	2.9515	2.8615	2.7965
	0	5	3.4568	3.3879	3.2907	3.1924	3.1092
	0.5	0.5	3.0140	2.9116	2.8210	2.7402	2.6828
	0.5	1	3.2383	3.1298	3.0265	2.9346	2.8647
	0.5	5	3.4606	3.3999	3.3068	3.2091	3.1264
	2	0.5	3.1988	3.1233	3.0364	2.9495	2.8830
	2	1	3.3337	3.2555	3.1613	3.0677	2.9926
	2	5	3.4659	3.4215	3.3372	3.2410	3.1601

3.2.2. Buckling Analysis of Multi-Directional Sandwich Plates

Likewise, a buckling analysis was carried out for a multi-directional FG skin paired with a homogeneous hard core under various boundary conditions, incorporating different values of p_x and p_z. Across all boundary conditions, the (1-2-1) scheme consistently demonstrated the highest frequency, followed by (2-2-1), (1-1-1), (2-1-2), and (1-0-1). The results are presented in Table 9.

The buckling load undergoes a notable decrease when p_x remains at zero and p_z varies from 0.5 to 5. Additionally, as both p_x and p_z are incrementally adjusted, the buckling load decreases, with p_z exerting a more significant influence than p_x. The maximum buckling load was observed under the FCFC boundary condition, followed by the CCCC, CSCS, and SSSS conditions.

The buckling analysis of a multi-directional FG skin with a homogeneous soft core was conducted using similar parameter values. The results are presented in Table 10. Across all boundary conditions, the (1-0-1) scheme consistently exhibits the highest buckling load, followed by (2-1-2), (1-1-1), (2-2-1), and (1-1-1). The buckling load experiences a significant increase when p_x is maintained at zero and p_z varies from 0.5 to 5. Furthermore, incremental adjustments in both p_x and p_z result in an increased buckling load. The maximum buckling load was observed under the FCFC condition, followed by the CCCC, CSCS, and SSSS conditions.

Table 9. Dimensionless buckling load $\overline{N}$ of square plates under various boundary conditions ($\xi_1 = 1$, $a/h = 10$). Type A: hard core.

Boundary Conditions	p_x	p_z	1-0-1	2-1-2	1-1-1	2-2-1	1-2-1
SSSS	0	0.5	3.6822	3.9704	4.2182	4.4049	4.6084
	0	1	2.5836	2.9200	3.2324	3.4747	3.7533
	0	5	1.3291	1.5213	1.7898	2.0560	2.3673
	0.5	0.5	3.3400	3.6008	3.8391	4.0419	4.2426
	0.5	1	2.3939	2.6942	2.9871	3.2338	3.5029
	0.5	5	1.3132	1.4882	1.7421	2.0049	2.3062
	2	0.5	2.6494	2.8539	3.0711	3.3002	3.5003
	2	1	2.0109	2.2382	2.4909	2.7428	2.9958
	2	5	1.2799	1.4213	1.6460	1.9017	2.1826
CSCS	0	0.5	6.8605	7.3953	7.8497	8.1872	8.5586
	0	1	4.8432	5.4737	6.0517	6.4943	7.0067
	0	5	2.5016	2.8785	3.3849	3.8795	4.4612
	0.5	0.5	6.2201	6.7143	7.1567	7.5246	7.8959
	0.5	1	4.4844	5.0537	5.5993	6.0509	6.5494
	0.5	5	2.4677	2.8154	3.2956	3.7841	4.3477
	2	0.5	4.9270	5.3365	5.7488	6.1675	6.5439
	2	1	3.7581	4.2053	4.6824	5.1459	5.6195
	2	5	2.3949	2.6880	3.1153	3.5909	4.1184
CCCC	0	0.5	9.2373	9.9552	10.5596	11.0035	11.4961
	0	1	6.5513	7.4037	8.1779	8.7647	9.4479
	0	5	3.3941	3.9213	4.6095	5.2737	6.0585
	0.5	0.5	8.3721	9.0459	9.6401	10.1254	10.6224
	0.5	1	6.0623	6.8391	7.5736	8.1735	8.8414
	0.5	5	3.3439	3.8349	4.4887	5.1449	5.9063
	2	0.5	6.6247	7.2047	7.7675	8.3230	8.8337
	2	1	5.0717	5.6980	6.3470	6.9651	7.6049
	2	5	3.2349	3.6606	4.2447	4.8841	5.5985
FCFC	0	0.5	10.8692	11.7121	12.4173	12.9312	13.5048
	0	1	7.7331	8.7391	9.6467	10.3296	11.1282
	0	5	4.0148	4.6515	5.4666	6.2465	7.1712
	0.5	0.5	9.8488	10.6484	11.3463	11.9092	12.4918
	0.5	1	7.1529	8.0755	8.9396	9.6388	10.4219
	0.5	5	3.9521	4.5488	5.3239	6.0947	6.9926
	2	0.5	7.7875	8.4932	9.1618	9.8086	10.4127
	2	1	5.9771	6.7339	7.5029	8.2251	8.9797
	2	5	3.8149	4.3412	5.0358	5.7873	6.6313

Table 10. Dimensionless buckling load $\overline{N}$ of square plates under various boundary conditions ($\xi_1 = 1$, $a/h = 10$). Type B: soft core.

Boundary Conditions	p_x	p_z	1-0-1	2-1-2	1-1-1	2-2-1	1-2-1
SSSS	0	0.5	3.9158	3.6018	3.3496	3.1017	2.9770
	0	1	4.9809	4.5812	4.2423	3.8888	3.7302
	0	5	6.3116	5.9908	5.6258	5.1429	4.9673
	0.5	0.5	4.2926	3.9860	3.7234	3.4429	3.3158
	0.5	1	5.2096	4.8331	4.4959	4.1236	3.9641
	0.5	5	6.3410	6.0394	5.6855	5.2028	5.0310
	2	0.5	5.0219	4.7482	4.4749	4.1298	3.9982
	2	1	5.6446	5.3254	5.0012	4.5931	4.4361
	2	5	6.3969	6.1343	5.8037	5.3219	5.1589

Table 10. *Cont.*

Boundary Conditions	p_x	p_z	1-0-1	2-1-2	1-1-1	2-2-1	1-2-1
CSCS	0	0.5	6.9651	6.3522	5.8856	5.4808	5.2319
	0	1	8.8733	8.0495	7.3940	6.8224	6.4704
	0	5	11.4549	10.6674	9.8643	9.0650	8.5544
	0.5	0.5	7.7135	7.0766	6.5604	6.0956	5.8111
	0.5	1	9.3552	8.5479	7.8692	7.2585	6.8763
	0.5	5	11.5328	10.7791	9.9902	9.1863	8.6741
	2	0.5	9.1297	8.5002	7.9139	7.3319	6.9766
	2	1	10.2468	9.5073	8.8092	8.1261	7.6953
	2	5	11.6792	10.9948	10.2374	9.4264	8.9138
CCCC	0	0.5	9.0716	8.2251	7.6020	7.1061	6.7586
	0	1	11.5691	10.3968	9.4986	8.8027	8.2856
	0	5	15.1311	13.9037	12.7229	11.7333	10.9025
	0.5	0.5	10.1175	9.2048	8.4892	7.9139	7.4925
	0.5	1	12.2651	11.0898	10.1375	9.3863	8.8056
	0.5	5	15.2571	14.0721	12.9036	11.9041	11.0635
	2	0.5	12.0741	11.1201	10.2665	9.5369	8.9691
	2	1	13.5375	12.4139	11.3971	10.5446	9.8545
	2	5	15.4932	14.3959	13.2576	12.2414	11.3854
FCFC	0	0.5	10.4404	9.4306	8.7022	8.1545	7.7373
	0	1	13.3239	11.9014	10.8357	10.0699	9.4331
	0	5	17.5759	16.0092	14.5512	13.4495	12.3755
	0.5	0.5	11.6978	10.5847	9.7294	9.0892	8.5673
	0.5	1	14.1765	12.7309	11.5854	10.7529	10.0252
	0.5	5	17.7400	16.2199	14.7711	13.6554	12.5642
	2	0.5	14.0361	12.8349	11.7854	10.9667	10.2366
	2	1	15.7264	14.3105	13.0605	12.1066	11.2195
	2	5	18.0477	16.6251	15.2018	14.0618	12.9413

4. Conclusions

The comprehensive investigation outlined in this paper provides pivotal insights into the buckling and free vibration behavior of multi-directional FG sandwich plates under a spectrum of boundary conditions. Our rigorous validation process and in-depth analyses offer a clear understanding of the materials' responses across various load applications. The consistency between the results of different shape function models underscores the reliability of our analytical approach. The implications of this research are particularly salient for design engineers and materials scientists focusing on the development of unidirectional and multi-directional FG sandwich panels custom-designed for specialized applications.

The primary conclusions drawn from this study can be summarized as follows:

1. The boundary condition of FCFC invariably results in the highest frequency and buckling load values when compared to other tested conditions such as CCCC, CSCS, and SSSS. Noteworthy is the observation that the transverse grading parameter p_z demonstrates a more significant effect than the longitudinal grading parameter p_x on these outcomes.
2. In the context of sandwich plates with a hard core, an increment in the values of both p_x and p_z is associated with a reduction in the plates' natural frequency and buckling load. This situation is reversed for materials with a soft core, where an increase in p_x and p_z corresponds to a decrease in their frequency and buckling load. This inverse relationship is due to the increased presence of ceramic constituents in the FG material, which are introduced as the grading parameters p_x and p_z rise, thereby enhancing the natural frequency due to their higher stiffness relative to metals.
3. Structural configurations that have a thicker core are shown to yield a higher stiffness. Specifically, for cores predominantly made of ceramic, enhancing the ceramic layer

thickness effectively introduces additional stiffness akin to rigid plates, which elevates their natural frequencies. In contrast, for metal-based core configurations, an increase in metal core thickness imparts greater flexibility to the structure, leading to a decrease in its natural frequencies.

The insights gleaned from our research extend the existing knowledge base and provide a robust foundation for the optimized design of FG sandwich plates, catering to the evolving demands of advanced engineering applications. We anticipate that our findings will spur further studies, potentially exploring even wider parameter spaces and boundary conditions to enrich our understanding of the structural applications of FG materials.

Author Contributions: Conceptualization, L.H. and V.P.; methodology, L.H., R.M. and H.A.A.; software, L.H. and R.M.; validation, L.H., V.P. and R.M.; formal analysis, L.H., V.P. and H.A.A.; investigation, V.P. and R.M.; resources, V.P. and R.M.; data curation, L.H. and V.P.; writing—original draft preparation, L.H. and R.M.; writing—review and editing, L.H., V.P. and H.A.A.; visualization, V.P. and R.M.; supervision, L.H., V.P. and H.A.A. All authors have read and agreed to the published version of the manuscript.

Funding: This research received no external funding.

Data Availability Statement: The raw data supporting the conclusions of this article will be made available by the authors on request.

Conflicts of Interest: The authors declare no conflicts of interest.

References

1. Saleh, B.; Jiang, J.; Ma, A.; Song, D.; Yang, D.; Xu, Q. Review on the Influence of Different Reinforcements on the Microstructure and Wear Behavior of Functionally Graded Aluminum Matrix Composites by Centrifugal Casting. *Met. Mater. Int.* **2020**, *26*, 933–960. [CrossRef]
2. Ghatage, P.S.; Kar, V.R.; Sudhagar, P.E. On the numerical modelling and analysis of multi-directional functionally graded composite structures: A review. *Compos. Struct.* **2020**, *236*, 111837. [CrossRef]
3. Kumar, S.; Mitra, A.; Roy, H. Large amplitude free vibration study of non-uniform plates with in-plane material inhomogeneity. *Proc. Inst. Mech. Eng. Part L J. Mater. Des. Appl.* **2018**, *232*, 371–387. [CrossRef]
4. Kumar, V.; Singh, S.; Saran, V.; Harsha, S. Exact solution for free vibration analysis of linearly varying thickness FGM plate using Galerkin-Vlasov's method. *Proc. Inst. Mech. Eng. Part L J. Mater. Des. Appl.* **2021**, *235*, 880–897. [CrossRef]
5. Du, D.; Sun, W.; Yan, X.; Xu, K. Free vibration analysis of rotating thin-walled cylindrical shells with hard coating based on Rayleigh-Ritz method. *Proc. Inst. Mech. Eng. Part G J. Aerosp. Eng.* **2021**, *235*, 1170–1186. [CrossRef]
6. Alipour, M.M.; Shariyat, M.; Shaban, M. A semi-analytical solution for free vibration of variable thickness two-directional-functionally graded plates on elastic foundations. *Int. J. Mech. Mater. Des.* **2010**, *6*, 293–304. [CrossRef]
7. Nie, G.; Zhong, Z. Axisymmetric bending of two-directional functionally graded circular and annular plates. *Acta Mech. Solida Sin.* **2007**, *20*, 289–295. [CrossRef]
8. Shariyat, M.; Jafari, R. Nonlinear low-velocity impact response analysis of a radially preloaded two-directional-functionally graded circular plate: A refined contact stiffness approach. *Compos. Part B Eng.* **2013**, *45*, 981–994. [CrossRef]
9. Ahlawat, N.; Lal, R. Buckling and Vibrations of Multi-directional Functionally Graded Circular Plate Resting on Elastic Foundation. *Procedia Eng.* **2016**, *144*, 85–93. [CrossRef]
10. Nguyen, D.K.; Nguyen, Q.H.; Tran, T.T.; Bui, V.T. Vibration of bi-dimensional functionally graded Timoshenko beams excited by a moving load. *Acta Mech.* **2017**, *228*, 141–155. [CrossRef]
11. Van Do, T.; Nguyen, D.K.; Duc, N.D.; Doan, D.H.; Bui, T.Q. Analysis of bi-directional functionally graded plates by FEM and a new third-order shear deformation plate theory. *Thin-Walled Struct.* **2017**, *119*, 687–699. [CrossRef]
12. Lieu, Q.X.; Lee, D.; Kang, J.; Lee, J. NURBS-based modeling and analysis for free vibration and buckling problems of in-plane bi-directional functionally graded plates. *Mech. Adv. Mater. Struct.* **2019**, *26*, 1064–1080. [CrossRef]
13. Bhattacharyya, M.; Kapuria, S.; Kumar, A.N. On the Stress to Strain Transfer Ratio and Elastic Deflection Behavior for Al/SiC Functionally Graded Material. *Mech. Adv. Mater. Struct.* **2007**, *14*, 295–302. [CrossRef]
14. Nakamura, T.; Wang, T.; Sampath, S. Determination of properties of graded materials by inverse analysis and instrumented indentation. *Acta Mater.* **2000**, *48*, 4293–4306. [CrossRef]
15. Yas, M.H.; Moloudi, N. Three-dimensional free vibration analysis of multi-directional functionally graded piezoelectric annular plates on elastic foundations via state space based differential quadrature method. *Appl. Math. Mech.* **2015**, *36*, 439–464. [CrossRef]
16. Nguyen-Ngoc, H.; Cuong-Le, T.; Nguyen, K.D.; Nguyen-Xuan, H.; Abdel-Wahab, M. Three-dimensional polyhedral finite element method for the analysis of multi-directional functionally graded solid shells. *Compos. Struct.* **2023**, *305*, 116538. [CrossRef]

17. Huang, B.; Zhao, G.; Ren, S.; Chen, W.; Han, W. Higher-order model with interlaminar stress continuity for multi-directional FG-GRC porous multilayer panels resting on elastic foundation. *Eng. Struct.* **2023**, *286*, 116074. [CrossRef]
18. Ramteke, P.M.; Sharma, N.; Dwivedi, M.; Das, S.K.; Uttarwar, C.R.; Panda, S.K. Theoretical thermoelastic frequency prediction of multi (uni/bi) directional graded porous panels and experimental verification. *Structures* **2023**, *54*, 618–630. [CrossRef]
19. Karamanli, A. Transient vibration analysis of strain gradient multi-directional functionally graded microplates under a moving concentrated load. *Compos. Struct.* **2023**, *308*, 116678. [CrossRef]
20. Thai, S. Optimization of multi-directional functionally graded plates in thermal environment based on 3D isogeometric analysis and adaptive-hybrid evolutionary firefly algorithm. *Thin-Walled Struct.* **2023**, *190*, 111000. [CrossRef]
21. Tang, Y.; Li, C.-L.; Yang, T. Application of the generalized differential quadrature method to study vibration and dynamic stability of tri-directional functionally graded beam under magneto-electro-elastic fields. *Eng. Anal. Bound. Elem.* **2023**, *146*, 808–823. [CrossRef]
22. Daikh, A.A.; Belarbi, M.-O.; Vinh, P.V.; Li, L.; Houari, M.S.A.; Eltaher, M.A. Vibration analysis of tri-directionally coated plate via thickness-stretching and microstructure-dependent modeling. *Mech. Res. Commun.* **2024**, *135*, 104221. [CrossRef]
23. Chami, G.M.B.; Kahil, A.; Hadji, L.; Madan, R.; Tounsi, A. Free vibration analysis of multi-directional porous functionally graded sandwich plates. *Steel Compos. Struct.* **2023**, *46*, 263–277. [CrossRef]
24. Singh, A.; Kumari, P. Three-Dimensional free vibration analysis of composite FGM rectangular plates with in-plane heterogeneity: An EKM solution. *Int. J. Mech. Sci.* **2020**, *180*, 105711. [CrossRef]
25. Singh, A.; Naskar, S.; Kumari, P.; Mukhopadhyay, T. Viscoelastic free vibration analysis of in-plane functionally graded orthotropic plates integrated with piezoelectric sensors: Time-dependent 3D analytical solutions. *Mech. Syst. Signal Process.* **2023**, *184*, 109636. [CrossRef]
26. Vaishali; Mukhopadhyay, T.; Kumar, R.R.; Dey, S. Probing the multi-physical probabilistic dynamics of a novel functional class of hybrid composite shells. *Compos. Struct.* **2021**, *262*, 113294. [CrossRef]
27. Malikan, M.; Eremeyev, V.A. A new hyperbolic-polynomial higher-order elasticity theory for mechanics of thick FGM beams with imperfection in the material composition. *Compos. Struct.* **2020**, *249*, 112486. [CrossRef]
28. Mallick, A.; Gangi Setti, S.; Sahu, R.K. Centrifugally cast functionally graded materials: Fabrication and challenges for probable automotive cylinder liner application. *Ceram. Int.* **2023**, *49*, 8649–8682. [CrossRef]
29. Verma, R.K.; Parganiha, D.; Chopkar, M. A review on fabrication and characteristics of functionally graded aluminum matrix composites fabricated by centrifugal casting method. *SN Appl. Sci.* **2021**, *3*, 227. [CrossRef]
30. Abdalla, H.M.A.; Casagrande, D. An Intrinsic Material Tailoring Approach for Functionally Graded Axisymmetric Hollow Bodies Under Plane Elasticity. *J. Elast.* **2021**, *144*, 15–32. [CrossRef]
31. Elkotb, H.H.; Mostafa, R.; Samad, A.A.A.; Enab, T.A. Manufacturing and Characterization of Functionally Graded Material Automotive Piston Using Centrifugal Casting Technique. *Solid State Phenom.* **2021**, *318*, 13–24.
32. Madan, R.; Bhowmick, S. Fabrication and microstructural characterization of Al-SiC based functionally graded disk. *Aircr. Eng. Aerosp. Technol.* **2023**, *95*, 292–301. [CrossRef]
33. Suresh, S.; Mortensen, A. Functionally graded metals and metal-ceramic composites: Part 2 Thermomechanical behaviour. *Int. Mater. Rev.* **1997**, *42*, 85–116. [CrossRef]
34. Loh, G.H.; Pei, E.; Harrison, D.; Monzón, M.D. An overview of functionally graded additive manufacturing. *Addit. Manuf.* **2018**, *23*, 34–44. [CrossRef]
35. Madan, R.; Bhowmick, S. Modeling of functionally graded materials to estimate effective thermo-mechanical properties. *World J. Eng.* **2022**, *19*, 291–301. [CrossRef]
36. Rizov, V. Delamination Analysis of Multilayered Functionally Graded Beams which Exhibit Non-linear Creep Behavior. *J. Appl. Comput. Mech.* **2023**, *9*, 935–944. [CrossRef]
37. Dastjerdi, S.; Tadi Beni, Y.; Malikan, M. A comprehensive study on nonlinear hygro-thermo-mechanical analysis of thick functionally graded porous rotating disk based on two quasi-three-dimensional theories. *Mech. Based Des. Struct. Mach.* **2022**, *50*, 3596–3625. [CrossRef]
38. Karami, B.; Ghayesh, M.H. Moving load excited dynamics of multi-layered imperfect microplates based on various micromechanical models. *Int. J. Eng. Sci.* **2024**, *197*, 104017. [CrossRef]
39. Karami, B.; Ghayesh, M.H. Vibration characteristics of sandwich microshells with porous functionally graded face sheets. *Int. J. Eng. Sci.* **2023**, *189*, 103884. [CrossRef]
40. Rezaei Farimani, M.; Mohadeszadeh, M. Thermo-elastic bending analysis of FGM rotating plate with axial grading and modified rule of mixture. *J. Braz. Soc. Mech. Sci. Eng.* **2017**, *39*, 299–307. [CrossRef]
41. Delale, F.; Erdogan, F. The Crack Problem for a Nonhomogeneous Plane. *J. Appl. Mech.* **1983**, *50*, 609–614. [CrossRef]
42. Sobhy, M. Buckling and free vibration of exponentially graded sandwich plates resting on elastic foundations under various boundary conditions. *Compos. Struct.* **2013**, *99*, 76–87. [CrossRef]
43. Li, Q.; Iu, V.P.; Kou, K.P. Three-dimensional vibration analysis of functionally graded material sandwich plates. *J. Sound Vib.* **2008**, *311*, 498–515. [CrossRef]

44. Zenkour, A.M. A comprehensive analysis of functionally graded sandwich plates: Part 2—Buckling and free vibration. *Int. J. Solids Struct.* **2005**, *42*, 5243–5258. [CrossRef]
45. Thai, H.-T.; Nguyen, T.-K.; Vo, T.P.; Lee, J. Analysis of functionally graded sandwich plates using a new first-order shear deformation theory. *Eur. J. Mech.—A/Solids* **2014**, *45*, 211–225. [CrossRef]

computation

MDPI

Article

Using Machine Learning Algorithms to Develop a Predictive Model for Computing the Maximum Deflection of Horizontally Curved Steel I-Beams

Elvis Ababu [1,*], George Markou [1,2,*] and Sarah Skorpen [1]

[1] Civil Engineering Department Hatfield Campus, University of Pretoria, Pretoria 0028, South Africa; sarah.skorpen@up.ac.za
[2] Civil Engineering Department, Neapolis University Pafos, Danais Avenue, Paphos 8042, Cyprus
* Correspondence: u17061785@tuks.co.za (E.A.); george.markou@up.ac.za (G.M.)

Abstract: Horizontally curved steel I-beams exhibit a complicated mechanical response as they experience a combination of bending, shear, and torsion, which varies based on the geometry of the beam at hand. The behaviour of these beams is therefore quite difficult to predict, as they can fail due to either flexure, shear, torsion, lateral torsional buckling, or a combination of these types of failure. This therefore necessitates the usage of complicated nonlinear analyses in order to accurately model their behaviour. Currently, little guidance is provided by international design standards in consideration of the serviceability limit states of horizontally curved steel I-beams. In this research, an experimentally validated dataset was created and was used to train numerous machine learning (ML) algorithms for predicting the midspan deflection at failure as well as the failure load of numerous horizontally curved steel I-beams. According to the experimental and numerical investigation, the deep artificial neural network model was found to be the most accurate when used to predict the validation dataset, where a mean absolute error of 6.4 mm (16.20%) was observed. This accuracy far surpassed that of Castigliano's second theorem, where the mean absolute error was found to be equal to 49.84 mm (126%). The deep artificial neural network was also capable of estimating the failure load with a mean absolute error of 30.43 kN (22.42%). This predictive model, which is the first of its kind in the international literature, can be used by professional engineers for the design of curved steel I-beams since it is currently the most accurate model ever developed.

Keywords: structural engineering; structural steel; curved beams; machine learning; finite element modelling

1. Introduction

Although curved beams have been in existence for centuries, minimal guidance has been available on the design of curved steel elements. The fabrication of curved steel beams began in the 19th century when steel members were cast in a curved profile or built up from components into a curved profile. Fabrication advanced, utilising three-roll pressing or induction bending in situations where residual stresses were particularly significant or smaller radii were required [1]. However, the majority of steel used for construction is formed through the use of roller bending, which is a cold process. When curving open sections using cold processes, the flanges exert a significant force on the web, which could lead to local buckling. Therefore, additional rolls are provided inside the tension flange for sections that are susceptible to local buckling (Figure 1).

Horizontally curved beams experience significant residual stresses due to the manner in which they are formed. A steel member rolled at room temperature experiences mechanical residual stresses (compressive and tensile) that reduce the ultimate strength of the member by increasing its flexibility, which subsequently causes a decrease in buckling strength [2]. Tensile residual stresses are particularly detrimental as they are often the cause of fatigue failure and stress corrosion cracking, whereas compressive residual stresses can be somewhat beneficial as they can mitigate the origination and propagation of cracks [3].

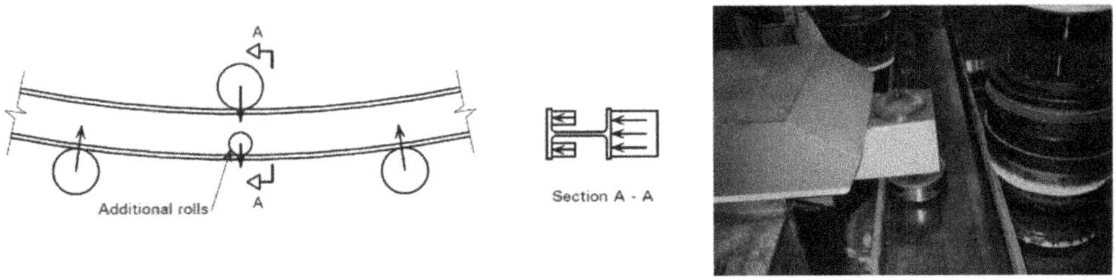

Figure 1. Additional rolls to prevent web buckling on an I-section [4].

The main challenge encountered with horizontally curved beams is that due to the geometry of the beam when normal loads are applied, the beam experiences a complex combination of bending, shear, and torsion simultaneously. Torsional moments can be visualised from the deflected shape as the compression flange tends to deflect laterally away from the centre of curvature [5]. When bending and torsional moments are applied simultaneously to a beam, the coupling of these forces tends to reduce the carrying capacity of the member. Research has found that the behaviour of horizontally curved beams is dependent on the *R/L* ratio, which can also be represented as the span angle. It has been noted that when the span angle is less than 1°, the beam responds similarly to a straight beam and is dominated by flexure (bending). When the span angle is larger than 20°, the beam behaviour of the beam is primarily dominated by torsion. When the span angle is between 1° and 20°, both bending and torsion significantly impact the behaviour of the beam [6].

In straight I-beams, lateral torsional buckling is easily observable through the lateral displacement and rotation of the member. In curved I-beams, however, this behaviour is always present due to the torsional moments experienced. This behaviour was termed lateral–torsional–vertical behaviour by Lee, et al. [7]. Therefore, due to initial curvature, similar to initial out-of-straightness in columns, bifurcation-type lateral torsional buckling may not be observed in curved steel I-beams. However, it has been noted that, as seen with straight beams, the flexural strength of curved beams decreases for members that are susceptible to lateral torsional buckling [6]. The effect of lateral torsional buckling is negligible when the angle between torsional restraints is less than 22.5°. Should this angle be larger than 22.5°, the American Institute of Steel Construction (AISC) recommends that the beam be handled as a straight beam with an adjusted lateral–torsional buckling modification factor that accounts for the curvature.

Over the years, numerous researchers have attempted to provide analytical formulae for horizontally curved members; however, very few have focused on equations regarding deflection and rotation. Wong [8] modified Castigliano's second theorem, which was originally developed for straight beams, and made it applicable to horizontally curved beams. Castigliano's second theorem is a strain energy method that is used to calculate deflection. The equation generated can be seen in Equation (1). C_d can be calculated from Figure 2. It is important to note here that Equation (1) is limited to beams that are fixed on both sides and a point load is applied at the centre point. Dahlberg [9] created various equations that can be used for various boundary and load conditions. However, these equations are quite difficult to implement and are therefore rarely used in practice. Furthermore, Dahlberg [9] did not compare the formula with experimental or finite element results.

$$\Delta_{0.5\theta} = \frac{Pr^3}{EI} C_d \tag{1}$$

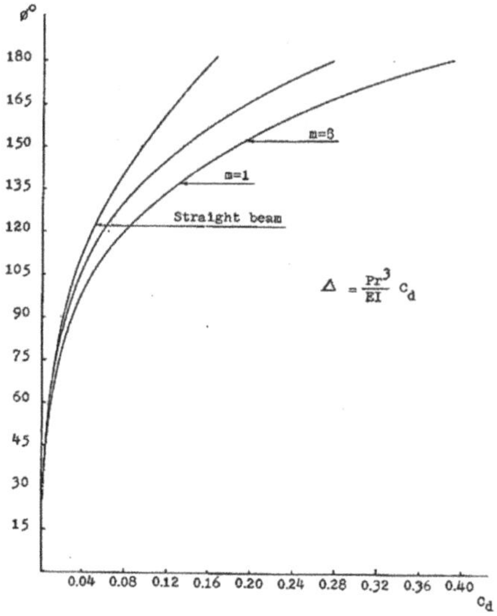

Figure 2. Variation of deflection coefficients with span angle for curved beams loaded with a concentrated load [8].

Castigliano's second theorem can also be applied to determine the rotation of curved beams; however, this is often even more tedious than the equations necessary for deflection. However, a simplified equation is provided by AISC (Equation (2)). The symbol h_0 represents the height of the member and Δ_{max} represents the maximum displacement of the member [6].

This equation is derived using the M/R method, which is used extensively in the design of horizontally curved beams, where the curved beam is modelled as a straight beam with a length equal to the curved member length. The torsional moment is then accounted for by a separate equation:

$$\theta_{max} = \tan^{-1}\left(\frac{2\Delta_{max}}{h_0}\right) \quad (2)$$

Various assumptions have been during the derivation of this formula, including that the thickness of each plate element is small relative to the width, which is, in turn, small relative to the span. The stresses due to warping are assumed to be negligible.

Modelling curved beams in FEM causes further complexities. The majority of commercially available software applications model curved beams as a series of straight beams, which provides sufficient accuracy for most design purposes. If the model experiences significant nonlinear behaviour, a convergence study is required in order to optimise the number of elements required, while the development of torsion further complicates the analysis process. Conventional beam finite elements (FEs) cannot be used to model horizontally curved steel I-beams that experience torsion, due to the existence of warping torsion. Conventional beam FEs account only for St Venant stiffness, which causes numerical error regarding computing the torsional deformation. Current beam FEs that incorporate warping stiffness are available; however, these are rarely incorporated in commercially available software and also tend to be less accurate. Three-dimensional FEs such as shell or solid elements are commonly used by researchers attempting to model the behaviour of curved beams. These elements accurately represent torsional behaviour without the need for numerous torsional constants. Previous researchers found that using solid FEs paired with nonlinear analyses led to accurate estimates of experimental results for both ultimate load and midspan deflection ([10–13]).

The combination of bending and shear as well as displacements and twisting rotations causes second-order bending actions in the plane of a curved I-beam. With an increase in load, these interactions grow rapidly and may ultimately cause early nonlinear behaviour and yielding, leading to reductions in the ultimate load-carrying capacity of the beam. This therefore necessitates the usage of nonlinear analyses when analysing horizontally curved beams, as opposed to general static (linear) analysis. Given the complexity of the problem, the use of FEs that are integrated with 3D numerical material models is also recommended, as in this research work.

2. Machine Learning Algorithms

This section briefly presents the various ways machine learning (ML) has been used to develop formulae that allow the design or mechanical investigation of civil engineering structures. ML and artificial intelligence (AI) have been implemented over recent decades in different fields as efficient tools used to predict analysis outputs for engineering problems that are deemed computationally demanding and highly complicated to solve analytically. The use of ML has largely eliminated the need for large numerical analysis, as it contains the ability to provide adequate estimates of the desired outputs. The main issue at hand, however, is that to train an ML algorithm, a large enough dataset is required. Therefore, the primary task of the majority of research work at this stage is the generation of datasets for various reinforced concrete and steel-related problems using either physical or validated numerical experiments. A meta-analysis of various case studies was presented by Markou, et al. [14]; however, the current paper focused on a single mechanical response problem in an attempt to provide a solution to a problem that has never been solved in the past. It is important to note that all ML algorithms that were used to perform the training, testing, and validation for the needs of this research work can be found on GitHub (https://github.com/nbakas/nbml/, accessed on 21 June 2023), through downloading nbml freeware. Furthermore, the datasets that were developed for the needs of this research work can be found through the following link (https://github.com/nbakas/nbml/tree/a0d27c94dd590688815180ebf6428963a24ca245/datasets, accessed on 1 July 2024), whereas the proposed models can be developed directly by the reader.

Various ML algorithms were used in this research, namely linear regression (LR), polynomial regression with hyperparameter tuning (POLYREG-HYT), hyperparameter tuning of extreme gradient boosting (XGBoost-HYT-CV), and parallel deep learning artificial neural networks with hyperparameter tuning (DANN-MPIH-HYT). The LR method was used as a point of comparison for the other ML algorithms (Markou, Bakas, Papadrakakis, and Chatzichristofis [14]).

POLYREG-HYT is useful for generating relatively accurate closed-form formulae and is applied to develop predictive models in higher-order classes. This relatively simplistic method provides a formula based on the nonlinear combination of all independent variables [14]. This model is based on the creation of nonlinear terms that are based on the independent variables up to the third degree. The algorithm then selects the nonlinear features that correspond to the minimum error. Originally, this methodology was utilised by Gravett, et al. [15], who made use of a simplistic approach in determining the number of features to use during training. Thereafter, that algorithm was improved and the use of hyperparameter tuning was introduced, as shown in Markou, Bakas, Papadrakakis, and Chatzichristofis [14], showcasing the proposed ML algorithm that has been used for the needs of this research work. The improved algorithm is outlined in Algorithm 1.

XGBoost-HYT-CV is a modification to the currently open-source extreme gradient boosting algorithm. The original XGBoost is a gradient-boosting library for ML problems such as classifying and regressing. The algorithm implements the gradient boosting framework, which is designed to be fast and scalable, making it suitable for large datasets [14]. For improving the current XGBoost algorithm, hyperparameter tuning was used. This tuning was found to exhibit accurate results compared to those of deep learning and require less computing demand [14]. One benefit of this algorithm is its ability to locate and replace missing values in the training and testing datasets.

Algorithm 1: Feature selection algorithm for polynomial regression [14]

```
Data: X, y, m_f (maximum number of features)
Result: Initialize [o] = 1 with the constant term ∈ [p]
Solve Linear System X' × a = y, where X' ⊂ X, with [o] columns.
Compute regression errors e_1.
Set as optimal error ê ← e_1.
Set as optimal indices [ô] ← [o].
for i ∈ [1,2,...,l] do
    repeat
        Select an index d ∈ [p] randomly.
        if d ∈ [o] then
            r ← U(0,1)
            if r < 1/2 then
                Select randomly o_d ∈ [p] : o_d ∉ [o]
                [o] ← ([o] \ d) ∪ o_d;
            else
                [o] ← [o] \ d;
            end
        else
            if o < m_f then
                [o] ← [o] ∪ d;
            else
                Select randomly o_d ∈ [o]
                [o] ← ([o] \ o_d) ∪ d;
            end
        end
    until rank(X') ≡ o;
    Solve Linear System X' × a = y.
    Compute regression error e_i.
    if e_i < ê then
        ê ← e_i
        [ô] ← [o]
    else
        [o] ← [ô]
    end
end
```

The final algorithm to be engaged was DANN-MPIH-HYT, which has the ability to train on both small and large datasets through the use of parallel processing. The algorithm was programmed to train on both computer processing units (CPUs) and graphics processing units (GPUs) and the use of distributed computations was also provided. This led to faster training and testing times when using the algorithm, as deep learning is known to be particularly slow when dealing with training processes, especially in cases where large datasets are used [14]. A combination of Horovod with MPI was used to optimise the numerical procedure. The Horovod library was implemented for multi-GPU training. Using Horovod, one is able to take a single GPU training script and run it across numerous GPUs in parallel. Through the use of message-passing interface (MPI) commands, each process is initialised and assigned its MPI rank in a straightforward manner, which is achieved in fewer code changes compared with other approaches. Various experimental algorithms were tested by Markou, Bakas, Papadrakakis and Chatzichristofis [14] on the Cyclone Supercomputer, utilising PyTorch for computer vision as well as regression tasks, highlighting the efficiency of data parallelism.

ML combined with FEM modelling has been used to create accurate formulae for various engineering applications in recent years. Markou and Bakas [16] created formulae to determine the shear capacity of concrete slender beams without stirrups. In their research work, a total of 35,849 beams were created and four ML algorithms were used, namely linear regression, polynomial regression, XGBoost, and deep learning neural networks. In that study, it was noted that the XGBoost algorithm was the most accurate, boasting an error of 5.82%. A similar methodology was followed by [17–20], where design formulae and predictive models were developed in relation to RC and steel structural problems. In this context, the current research work validates numerical models through the use of results from experiments conducted on curved steel I-beams at the laboratories of the University of Pretoria, and thereafter, the development of a relatively large dataset through nonlinear analyses is described. The next step will involve the development of the proposed predictive models that are validated and presented in the present paper.

3. Experimental Investigation

Experimental studies were performed in order to validate the finite element models that were used to develop the datasets for the needs of this research work. Experiments were also used to determine the appropriate finite element type that would yield the most accurate results regarding the deflection and rotation of the I-beam. The experiments involved the investigation of a horizontally curved steel I-beam (IPE 100) where the two ends were fully fixed and a vertically upward load was applied at the midspan. The results measured were vertical deflection at the midspan, rotation at the midspan and quarter points, and strains at the midspan and the support.

The beam that was analysed was a 3.5 m long IPE100 beam. This beam was selected to minimise the applied forces in order to maintain a safe working environment. A simplified schematic outlining how the beam was supported and loaded can be seen in Figure 3.

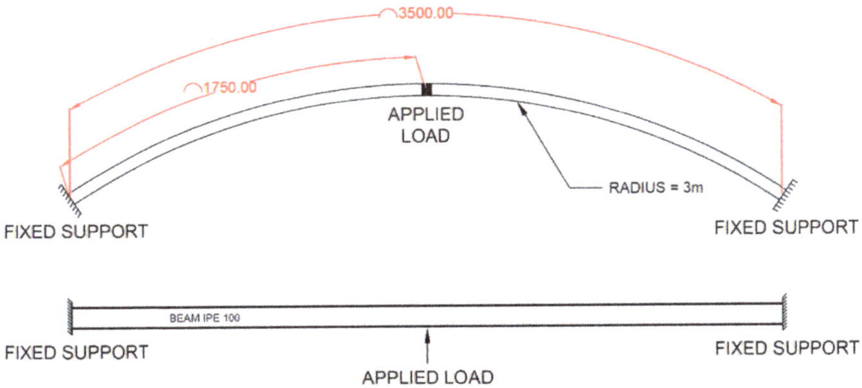

Figure 3. Basic schematic of the curved beam experimental setup (all dimensions are in millimetres).

To create a fully fixed support, the cross-section of the IPE 100 beam was welded onto two 12 mm thick plates. These 12 mm thick plates were then welded onto a stiff beam, which was then bolted onto the test floor using M24 Gr8.8 bolts. The entire system was assumed rigid and minimal deflections and rotations were expected at this point. Additional linear variable differential transformers (LVDTs) were placed at the supports. A schematic of the support conditions as well as a photograph can be seen in Figures 4 and 5, respectively.

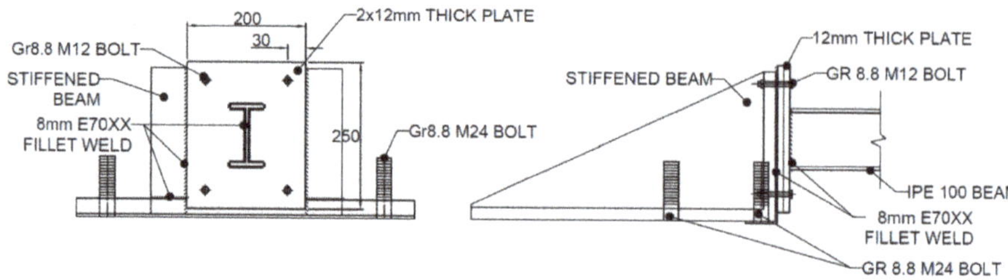

Figure 4. Schematic of the experiment with fixed support (all dimensions are in millimetres).

Figure 5. Experiment with fixed support.

A simple load application mechanism was used, which allowed a quicker and easier test set-up. The beam was loaded with a 40 mm steel bar that had a flat edge. This flat edge was loaded onto the beam using an overhead crane. The load cell then calculated the load experienced by the beam every 100 milliseconds, which was seen to provide sufficient data for analysis purposes. The loading equipment had a capacity of 50 tonnes, which was sufficient for the expected failure load of the beam. The steel bar also allowed the beam to rotate freely at the point of load application, which was required given the large expected rotations of this horizontally curved beam.

LVDTs were used to measure beam deflection only in the vertical direction. Two different LVDT types were used, one with a 250 mm range, which was placed at the supports and the other with a 1000 mm range, which was placed at the midspan. The LVDT at the midspan was placed 160 mm away from the point of load application so measurements could be taken at the same position as the strain gauges. The strain gauges were offset in order to not be influenced by the applied point load. LVDTs were used due to their high accuracy and good long-term stability as opposed to potentiometers, which have lower accuracy and precision but are typically more versatile. A schematic of the placement of LVDTs can be seen in Figure 6.

Inclinometers were used to measure the rotation along the beam as well as across the beam. Two dual-axis inclinometers were used. The inclinometers used made use of an electrolytic level which was capable of measuring inclination along two axes (pitch and roll). One inclinometer was placed 100 mm from the point of load application but on the opposite half of the beam from where the LVDT and strain gauges were placed, and the other was placed at the quarter point of the same half. A schematic outlining the position of the inclinometers can be seen in Figure 7.

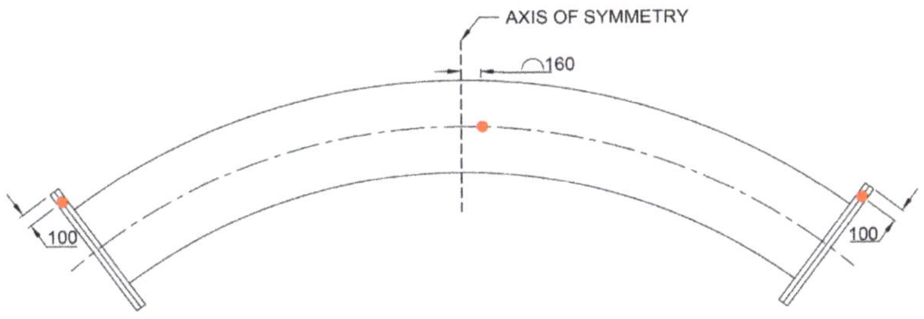

Figure 6. Schematic indicating points of displacement measurements (all dimensions are in millimetres).

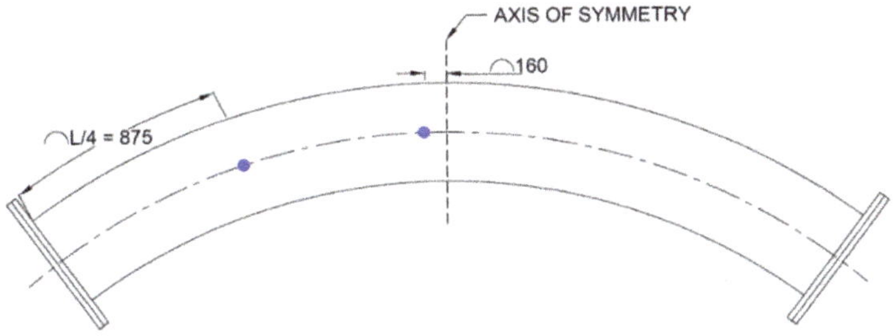

Figure 7. Schematic indicating points of rotation measurements (all dimensions are in millimetres).

The majority of the setup of the instrumentation involved accurately placing the strain gauges. Strain gauges are very sensitive. Therefore, the surface where a strain gauge is applied is required to be extremely smooth and free of all impurities prior to the placement of the strain gauge. The placement of these strain gauges was critical so that comparison between the strain gauge and the FE analysis could be as accurate as possible. This was particularly true for the strain gauges placed at 45°, which were required to accurately measure torsional strain within the beam. A total of 16 strain gauges were used to measure the strain at various points on one half of the beam. At the midspan, only longitudinal strain gauges were provided at the top and the bottom flange. These longitudinal flanges were placed at the far edge of the beam to read the maximum longitudinal strain at the midspan. The same setup was followed at the support, which led to another four longitudinally placed strain gauges being placed at this point. Two strain gauges were also placed transversely at the top and bottom flanges to investigate the shear flow in the beam. A further two vertical strain gauges were placed on either side of the web to measure the shear flow of the beam across the web. Four diagonal strain gauges were placed to measure the torsional stresses experienced at these points. A schematic outlining the position of the strain gauges can be seen in Figure 8.

All instrumentation was connected to an HBX Quantum logger. This logger allowed conversion from voltage to the appropriate units for the various instrumentation after calibration (e.g., millimetres for the LVDT, kilonewtons for the load cell, and micrometres/metres for the strain gauges). During the experiment, the load was applied to the beam in increments up to the final failure load past the point of yielding. During loading, a video was taken to record the experiment, which was used for verification when analysing the results. Photos were then also taken after the final load was applied and yielding had been experienced for further analysis (Figure 9).

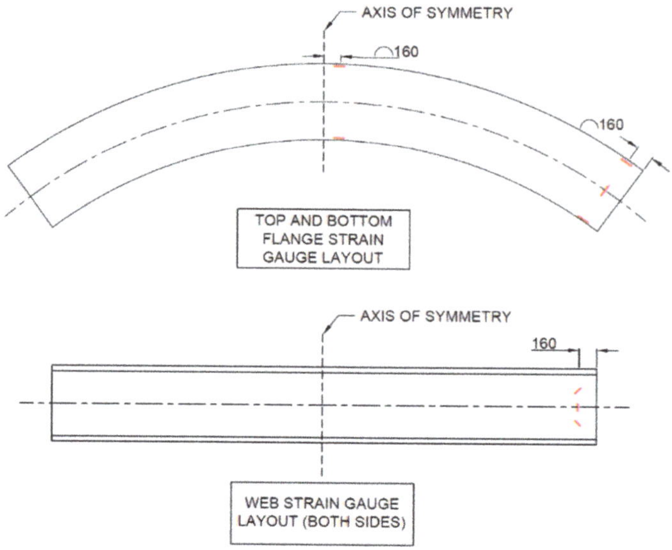

Figure 8. Schematic indicating points of strain measurements (all dimensions are in millimetres).

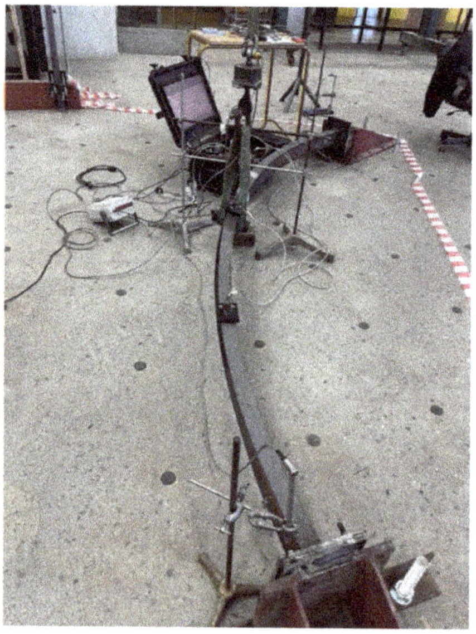

Figure 9. Deflected shape of the beam during the loading phase.

4. Analysis of Experimental Results

During testing, a total of five variables were measured. These variables were the load applied at the midspan, the displacement, the tilt/rotation, and the strain. The initial readings prior to the load application were taken as a reference point for all future readings. As previously mentioned, the load was applied incrementally in steps of 3 kN. This was carried out so it could be accurately determined at what load yielding occurred (through measuring permanent deformation). It was noted that after the 12 kN load was removed, permanent deformation occurred at the midpoint of the beam. The deflection for this load was approximately 55 mm.

The deflection measured at the supports was negligible, reaching up to 0.07 mm (upwards). This was seen as negligible and therefore was not factored into the calculations of the midspan deflection. Figure 10 shows the load–deflection curve of the beam. Results showed that the beam deflected linearly up to 55.05 mm at a load of 12.42 kN. The experiment was stopped right at the point of yielding; therefore, a plateau is not indicated in this graph, which would be a further indication that yielding had occurred.

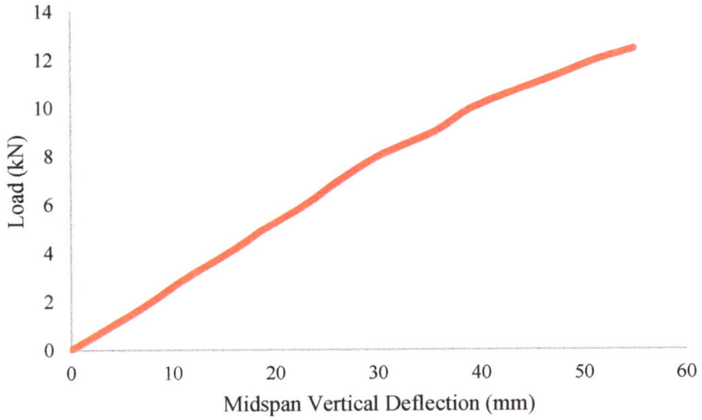

Figure 10. Experimental beam load–deflection curve (midspan).

When investigating the tilt results, errors were noted. The results showed that tilt did not consistently increase as load increased. At some points, a decrease in tilt was measured with an increase in load. This can be seen in Figure 11, where it can be seen that as the load increased from 2 kN to 4 kN, the tilt was measured to have decreased slightly. Practically, this is not possible. It is believed that this error was due to incorrect use of the measuring apparatus. The apparatus used to measure tilt made use of an electrolytic level. Electrolytic levels are typically poor when it comes to dynamic/cyclical loading, due to the conductive fluid found within the sealed glass. These results can therefore be used to obtain a rough estimate of how tilt increases as the load increases; however, these results cannot be used to determine the accuracy of FE models or analytical formulae. A line of best fit has been included in Figure 11 for further interpretation, which shows that tilt increased as load increased.

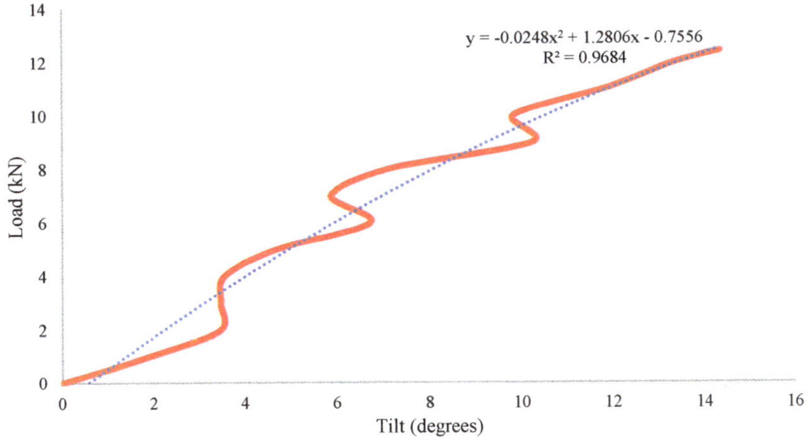

Figure 11. Experimental beam load–tilt diagram.

The strain results were quite interesting and showed the implication of the support conditions utilised. The support was intended to be fully fixed; however, practically, the support behaved in a way between that of a fully fixed connection and a torsionally pinned connection, as was seen with the strain gauge results. The strain gauge pattern was symmetrical with the cross section (both vertically and horizontally); however, the results were not seen to be perfectly symmetrical. The major reason for the discrepancies was most likely to have been due to inaccuracies during the setting out of the strain gauges. Material and geometric imperfections also played a role. The strain gauge results can be seen from Figures 12–14.

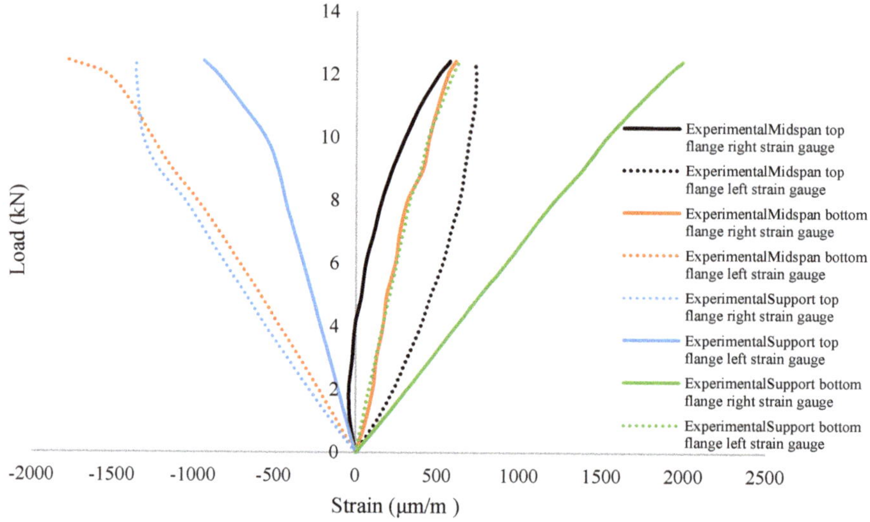

Figure 12. Graphical summary of experimental beam flange longitudinal strain gauge results.

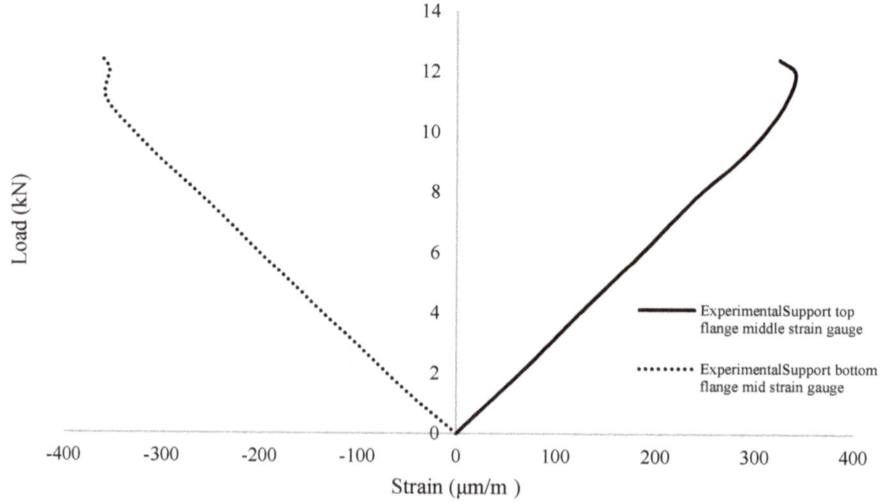

Figure 13. Graphical summary of experimental beam flange transverse strain gauge results.

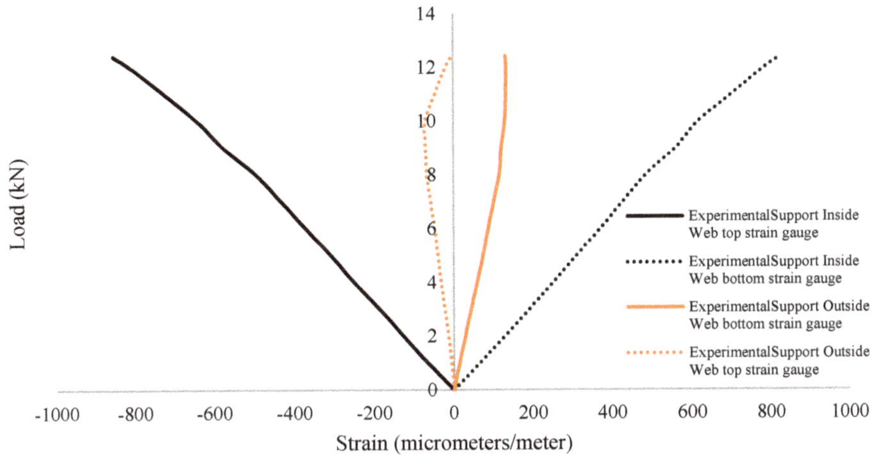

Figure 14. Graphical summary of experimental beam 45° web strain gauge results.

5. Comparison of Experimental Data with Finite Element Analysis Results

Various FE models were used and compared to determine which model most accurately represented the behaviour of the experimental beam. The focus was to determine the optimum FE model to be used for the development of the datasets. The datasets were used to train various ML algorithms for the development of the proposed predictive models related to the calculation of the ultimate strength of curved steel I-beams that are fixed at their ends and the respective deflection.

CivilFEM 2021 was used to develop the numerous FE meshes that were used to construct the dataset. This software was selected due to its interface, which allowed the creation of curved steel I-beam meshes in an automated manner. Solid and shell FEs were considered in this research work, where the mesh size was determined based on a previous research study where a mesh sensitivity analysis was conducted by the authors [21]. Linear and quadratic elements were also considered for each of the FEs used to reproduce the experimental results. It is important to note at this stage that [21] used experimental results found in the literature to calibrate their model that involved steel I-beams with a different support system compared with the fixed-end beams that are investigated herein. That pilot project showed that hexahedral isoparametric FEs were able to accurately reproduce experimental results [21].

Therefore, a total of four models were created for the needs of this research work, namely, a linear solid element, a quadratic solid element, a linear shell element, and a quadratic shell element. All material-related parameters were kept constant between the four models to maintain consistency. A typical nonlinear material model with anticipated bilinear behaviour was assumed. The hardening type selected was isotropic, indicating that yielding occurred when the effective or equivalent stress was equal to the specified yield stress. This selected yielding criterion was set to be that of von Mises. Furthermore, the material properties were assumed to be that of an S355 steel material (355 MPa yield stress, 210 GPa elastic modulus, a Poisson ratio of 0.3, and a density of 7850 kg/m^3). The experimental deflection and rotation at the midspan were then compared with the midspan deflection and rotation calculated using the various FE models. Analytical formulae were also included for comparative purposes.

Figure 15 shows that the linear solid FE derived the worst prediction when estimating midspan deflection. The analytical method, Castigliano's second theorem, was also unable to accurately represent the behaviour of the beam and performed quite poorly compared with the more advanced FE models. The quadratic solid model most accurately calculated the midspan deflection, with an average error of only 1.83 mm. A summary of the computed mean absolute error (MAE) can be seen in Table 1.

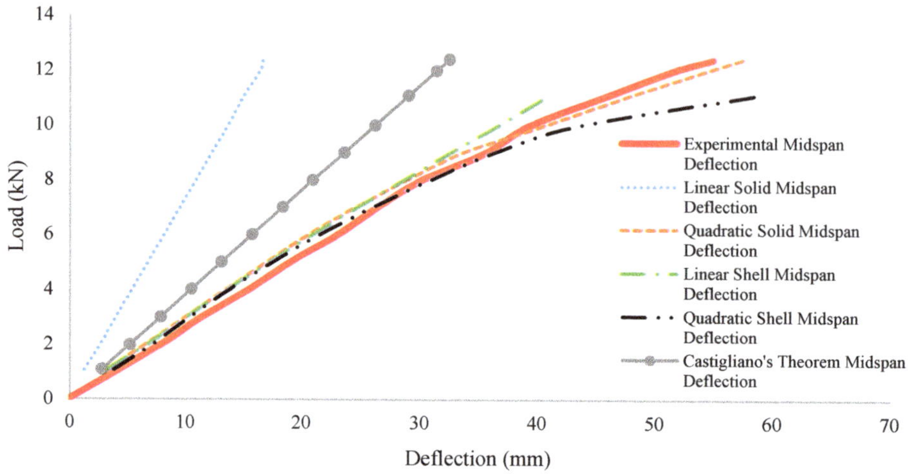

Figure 15. Load–deflection graph displaying experimental deflection results, FE results, and results obtained with the analytical formula.

Table 1. Mean absolute error of the FEM deflection estimate and Castigliano's theorem compared with the experimental data.

Deflection Estimation Method	MAE
Linear solid finite element model	18.64 mm
Quadratic solid finite element model	1.83 mm
Linear shell finite element model	2.12 mm
Quadratic shell finite element model	2.33 mm
Castigliano's theorem	9.91 mm

Furthermore, the resulting section rotations can be seen in Figure 16. Similar to the deflection results, experimental data and the results derived from FE analyses were compared with those of a commonly used analytical method, as discussed previously (Equation (2)). The experimental deflection was used as an input to achieve the highest accuracy possible when implementing the analytical method. Due to the fact that solid FEs do not contain rotational degrees of freedom, the rotation was calculated from first principles through making use of the vertical deflection at opposite ends of the bottom flange and using trigonometry to calculate the rotation. This method only holds given that the cross-section does not warp significantly. From the graphic, it was noted that the linear solid FEs provided the least accuracy, underestimating the midspan tilt quite significantly.

The M/R performed slightly better compared with the linear solid Fes; however, these results were still exceedingly conservative compared with the more advanced FE models. The more advanced FE models correlated well with the line of best fit of the experimental data. However, it seems the quadratic shell model was more accurate up until approximately 10 kN. From 10 kN to failure, the linear shell and the quadratic solid models both displayed lower errors compared with the experimental data. A summary of error metrics (MAE) can be seen in Table 2.

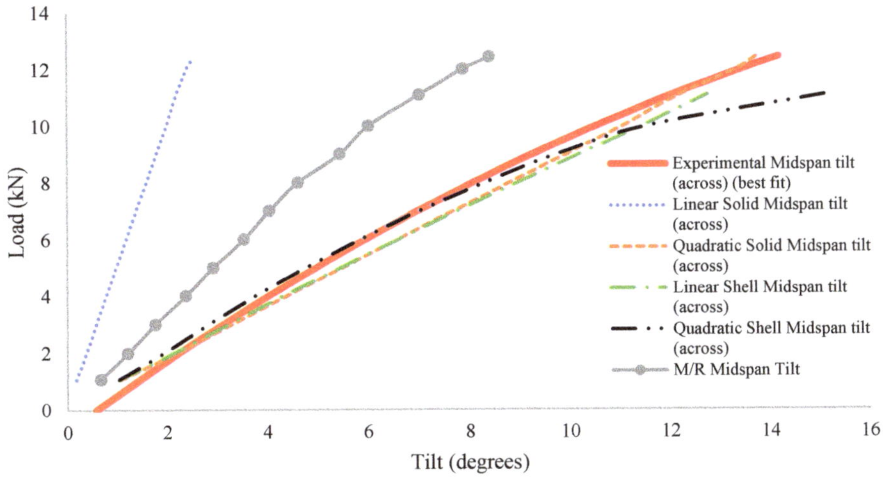

Figure 16. Load–rotation graph displaying experimental, FE, and the M/R analytical formula results.

Table 2. Mean absolute error of FEM rotation estimate and the analytical formula for experimental data.

Rotation Estimation Method	MAE
Linear solid finite element model	4.98°
Quadratic solid finite element model	1.31°
Linear shell finite element model	1.52°
Quadratic shell finite element model	1.33°
M/R method	2.31°

It is therefore concluded that FE models with the numerical modelling technique as outlined in this section are capable of accurately representing the behaviour of horizontally curved steel I-beams. These models have also been shown to outperform quite significantly the existing analytical formulae for estimating both deflection and rotation. It was noted that of all the FE models, the quadratic solid FE performed best in both estimating deflection and rotation. Therefore, the numerical modelling technique outlined in this section was utilised in developing a large dataset as described in the subsequent sections.

6. Numerical Campaign

A relatively large dataset consisting of 864 models was created through the use of CivilFEM. A total of six sections were considered, namely, IPE100, IPE200, IPE300, IPE400, IPE500, and IPE600. This was done to encompass the entire IPE section list according to European standards. The variables considered were the moment of inertia around the x axis (I_{xx}), the moment of inertia around the y axis (I_{yy}), polar moment of inertia (J), section height (h), section width (b), section area (A), beam curved length (L), beam radius of curvature (R), material yield strength (f_y), material Young's modulus (E), material Poisson ratio (v), and material shear modulus (G). For each section considered, three yield strengths were assumed (235 MPa, 275 MPa, and 355 MPa), to account for the various structural steel strengths commercially available in South Africa. Three Young's moduli were considered, namely, 190 GPa, 200 GPa, and 210 GPa, to account for variances in material stiffness. The Poisson ratio was fixed at 0.28 and the equation of G was calculated based on E and v. Four lengths and four radii were considered per section in order to cover the broad range of geometries that the proposed predictive models will be applicable for. Given that R/L ratios have been found to control the behaviour of curved beams, four specific R/L ratios were considered on all beams, namely, $R/L = 1, 2, 4$, and 8. All the beams were fully fixed on both ends and a load was applied normally downwards at the midspan.

To maintain consistency, the models were automatically generated using a Python script within CivilFEM. This ensured all material and geometrical properties were maintained and only a single variable was modified, allowing automatic mesh generation. A Python script was developed and optimised to ensure consistency and minimise running time through decreasing the number of variables and the overall performance of the code. The models all made use of quadratic solid elements with a mesh size of up to 50 mm. A typical nonlinear material model with predicted bilinear behaviour was assumed. The hardening type selected was isotropic, which indicates that yielding occurred when the effective or equivalent stress was equal to the specified yield stress. This selected yielding criterion was set to be that of von Mises.

Table 3 indicates different statistical parameters, including skewness. A value of 0 indicates that the dataset contained no skewness and had a perfectly normal distribution. According to Aminu and Shariff [22], a range from −3 to 3 can be considered as a cutoff, and based on this, the results show that the data were not significantly skewed. Kurtosis is another parameter that assists further in determining whether a dataset is normally distributed or not. This is performed by determining whether a dataset is "heavy-tailed" or "light-tailed". "Heavy-tailed" implies that a dataset contains numerous data points in outlier positions, whereas a "light-tailed" dataset contains minimal to no outliers. Once again, there is no accepted convention on what is deemed "heavy-tailed"; however, Aminu and Shariff [22] state that a range from +10 to −10 is deemed light-tailed and contains minimal outliers. In this context, it was observed that the dataset created for the needs of this research work contained minimal outliers with a good distribution.

Table 3. Horizontally curved beam dataset descriptive statistics.

	Mean	Median	STD	Min	Max	Skewness	Kurtosis
I_{xx} (mm^4)	291,198,612	231,300,000	326,514,119	1,710,000	920,800,000	0.977	−0.455
I_{yy} (mm^4)	12,720,819	13,180,000	1,1960,328	159,000	33,870,000	0.625	−0.960
J (mm^4)	692,957	841,000	553,277	12,100	1,540,000	0.142	−1.343
h (mm)	349.80	400	172.05	100	600	−0.011	−1.284
b (mm)	150.59	180	58.02	55	220	−0.472	−1.181
A (mm^2)	7490	8450	5048	1030	15,600	0.285	−1.223
L (m)	7.48	4	6.44	1	24	1.346	0.953
R (m)	27.99	16	35.26	1	192	2.653	8.242
F_y (MPa)	289.37	275	50.23	235	355	0.343	−1.543
E (GPa)	199.93	200	8.14	190	210	0.012	−1.491

The correlation matrix for the displacement dataset can be found in Figure 17a and for the failure load dataset in Figure 17b. It was clear from the correlation matrix results that all cross-sectional properties had a strong positive correlation with the midspan deflection and the failure load. It was also clear that, of all the properties, the curved length had the largest correlation with the midspan vertical deflection. This implies the potential to translate to positive results in the sensitivity analysis; this is discussed in Section 7.

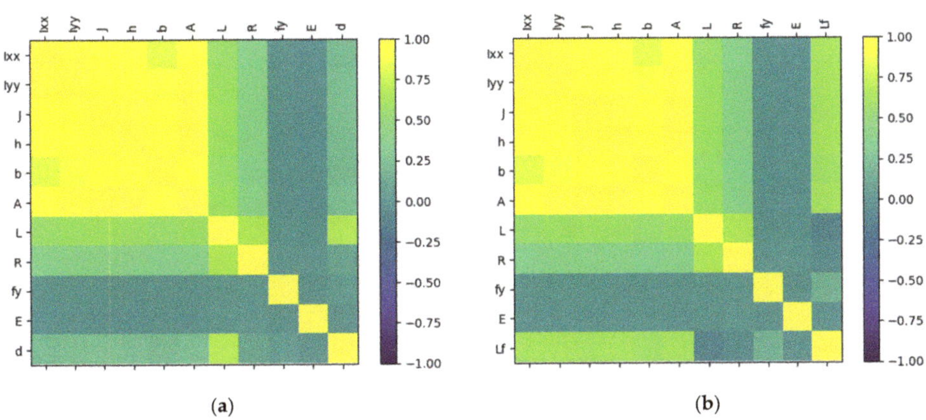

Figure 17. Correlation matrix of (**a**) midspan deflection dataset and (**b**) failure load dataset.

7. Machine Learning Training and Testing

This section outlines how numerous ML algorithms were applied in order to create formulae that can outperform current analytical methods used to estimate the deflection of horizontally curved steel I-beams. The ML algorithms considered were linear regression (LR), polynomial regression with hyperparameter tuning (POLYREG-HYT), deep artificial neural networks with MPI, Horovod, and hyperparameter tuning (DANN-MPIH-HYT), and extreme gradient boosting with hyperparameter tuning and cross-validation (XGBoost-HYT-CV) [14]. It must be noted here that for the ML analyses performed, 85% of the dataset was used to train the ML algorithm and 15% was used for testing. The performance of ML algorithms varies; therefore, to quantify accuracy, numerous error metrics were considered. The error metrics considered in this research work were the mean absolute percentage error (MAPE), the mean absolute mean percentage error (MAMPE), the mean absolute error (MAE), and the root-mean-square error (RMSE). The Pearson correlation coefficient was also considered (R) for determining the similarity between the PV (predicted value) and the DV (dependent variable). An in-depth discussion of the error metrics can be found in [23].

7.1. Proposed Predictive Models for the Case of Deflection

This section outlines the midspan deflection predictions provided by the various ML algorithms. The numerically obtained error metrics were analysed to determine the performance of the various proposed predictive models. The correlation for the train and test datasets can be seen in Figures 18–21.

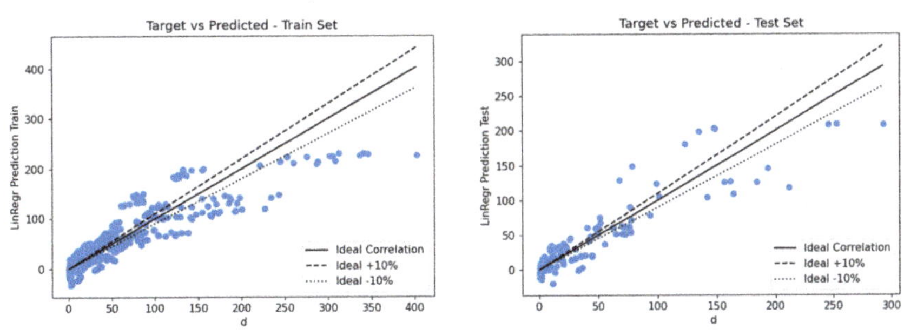

Figure 18. Correlation between the deflection determined using FEM and the deflection estimated using LR (**a**) on the training dataset and (**b**) on the testing dataset.

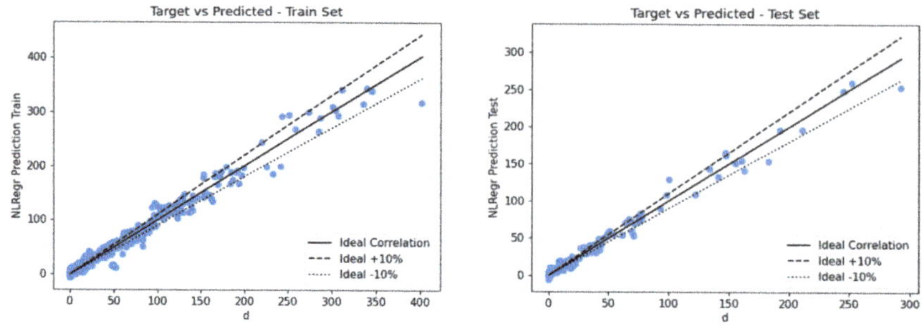

(**a**) (**b**)

Figure 19. Correlation between the deflection determined using FEM and the deflection estimated using POLYREG-HYT (**a**) on the training dataset and (**b**) on the testing dataset.

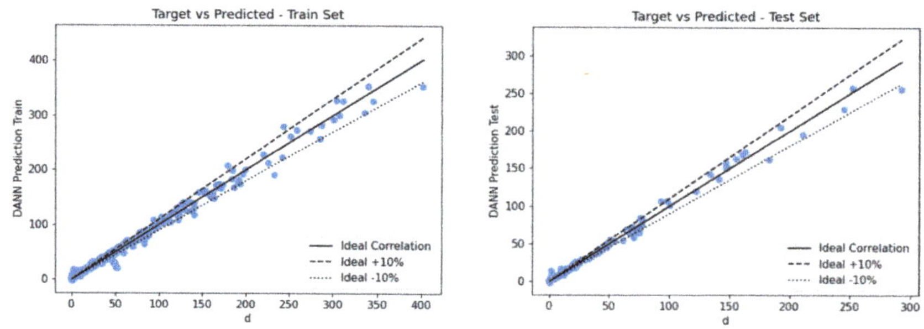

(**a**) (**b**)

Figure 20. Correlation between the deflection determined using FEM and the deflection estimated using the DANN-MPIH-HYT algorithm (**a**) on the training dataset (**b**) on the testing.

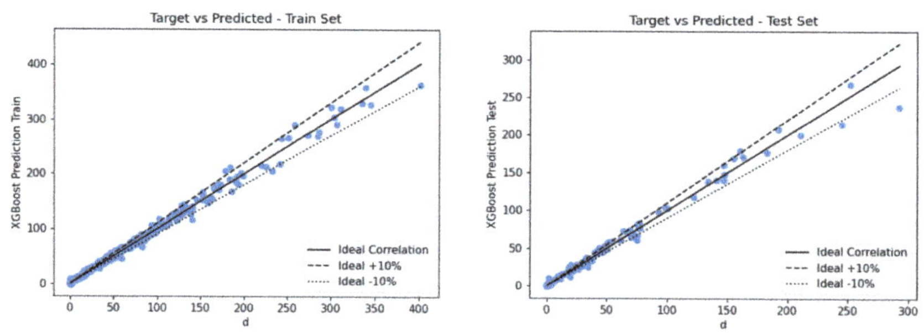

(**a**) (**b**)

Figure 21. Correlation between the deflection determined using FEM and the deflection estimated using the XGBoost-HYT-CV algorithm (**a**) on the training dataset (**b**) on the testing dataset.

LR and POLYREG-HYT are the only ML algorithms considered in this study that are capable of providing closed-form formulae. Equation (3) shows the formula generated with the LR algorithm, which can be used to estimate the vertical midspan deflection, whereas Equation (4) provides the proposed predictive model derived using POLYREG-HYT. When implementing the formulae, all cross-sectional properties are in mm/mm^2/mm^3/mm^4, and the span is given in m. The yield strength is provided in MPa and the Young's modulus in GPa.

$$d = -2.9147 \times 10^{-8}E - 08 \times I_{xx} + 6.085 \times 10^{-6} \times I_{yy} + 7.153 \times J + 8.6563 \times 10^{-1} \times h - 6.7409 \times 10^{-1} \times b - 4.496 \times 10^{-2} \times A + 1.1562 \times 10^{1} \times L - 1.1261 \times R + 5.1585 \times 10^{-2} \times f_y - 6.87948 \times 10^{-2}E \quad (3)$$

$$d = -1.10850 \times 10^{-2} \times h \times L + 5.54753 \times L^2 - 2.59341 \times 10^{-2} \times b \times L^2 + 8.71342 \times 10^{-1} \times L + 4.36940 \times 10^{-1} \times R - 3.72902 \times 10^{-4} \times b^2 + 7.64167 \times 10^{-5} \times h \times R^2 + 4.14113 \times 10^{-7} \times h \times A \times L + 1.20492 \times 10^{-1} \times R^2 - 2.10080 \times 10^{-9} \times J \times L \times f_y + 2.36217 \times 10^{-8} \times A^2 \times R + 8.06401 \times 10^{-3} \times b \times L \times R - 3.01987 \times 10^{-5} \times A \times L \times R - 1.61891 \times L \times R - 2.78160 \times 10^{-2} \times L^3 - 3.98564 \times 10^{-4} \times R^3 - 4.35591 \times 10^{-9} \times b \times A^2 - 6.98688 \times 10^{-4} \times b \times R^2 + 4.76077 \times 10^{-3} \times L \times R^2 + 9.28207 \times 10^{-12} \times I_{yy} \times b \times f_y - 2.05690 \times 10^{-7} \times A \times R \times f_y + 2.68695 \times 10^{-7} \times h^3 + 1.26135 \times 10^{-10} \times I_{yy} \times R^2 + 2.50786 \times 10^{-5} \times R^2 \times f_y + 8.68248 \times 10^{-4} \times L^2 \times f_y + 6.78638 \times 10^{-11} \times I_y \times R \times f_y - 1.03722 \times 10^{-6} \times h \times b \times f_y + 1.76647 \times 10^{-5} \times h \times b \times R - 3.55305 \times 10^{-7} \times A \times E^2 - 9.13425 \times 10^{-7} \times h \times {f_y}^2 + 1.87704 \times 10^{-7} \times h \times A \times E - 8.00661 \times 10^{-7} \times h \times A \times R + 4.02981 \times 10^{-6} \times h \times f_y \times E - 2.62557 \times 10^{-4} \times L \times R \times f_y - 1.01188 \times 10^{-2} \times L^2 + 1.18944 \times 10^{-6} \times J \times L^2 - 7.59868 \times 10^{-10} \times I_{yy} \times L \times E \quad (4)$$

Visually, it was difficult to discern which ML algorithm performed the best. Therefore, as previously mentioned, numerous error metrics were used to measure the performance of the various algorithms. Table 4 summarises the results. As can be seen, XGBoost-HYT-CV was the most accurate proposed model when looking at the training dataset, whereas when comparing the error metrics obtained from the testing dataset, that was not the case. The DANN-MPIH-HYT proposed predictive model was found to be slightly more accurate. It should also be noted that, even though DANN-MPIH-HYT was the most accurate in the testing phase, this algorithm did require the longest computation time (27 times slower).

Table 4. Comparison of the ML algorithms trained on the deflection dataset.

Model	Dataset	R	MAPE	MAMPE	MAE	RMSE	Computation Time (s)
LR	Train	0.8934	225.93%	45.08%	17.6614	25.6709	0.0013
POLYREG-HYT	Train	0.9874	69.91%	13.90%	5.4447	9.0447	10.91
DANN-MPIH-HYT	Train	0.9950	39.64%	7.59%	2.9731	5.7272	2019.76
XGBoost-HYT-CV	Train	0.9968	22.97%	5.92%	2.3199	4.5566	75.27
LR	Test	0.9064	168.22%	39.74%	17.1810	24.4716	0.00
POLYREG-HYT	Test	0.9905	70.38%	12.49%	5.3997	7.9865	0.00
DANN-MPIH-HYT	Test	0.9950	35.82%	8.04%	3.4776	5.8759	0.0004
XGBoost-HYT-CV	Test	0.9922	23.89%	8.44%	3.6504	7.2498	0.0033

Ultimately, it was apparent that accurate formulae can be created using ML algorithms. A later section of this paper compares these results with out-of-sample data, further validating the proposed predictive models' ability to capture unknown results.

Sensitivity analyses were conducted on each of the independent variables to determine which variable affected the dependent variable the most. This will assist future studies in determining which variables to exclude in order to increase efficiency in the proposed predictive models. This analysis also provides engineers with a deeper understanding of the behaviour of horizontally curved steel I-beams. Figure 22 summarises the findings of the sensitivity analysis. As can be seen, the three most influential variables on midspan vertical deflection were the curved length (*L*), the moment of inertia around the minor axis (I_{yy}), and the radius of curvature (*R*). *L* was by far the most influential variable, according to the obtained results. The cross-section area, cross-section width, and cross-section height had practically no impact on the midspan deflection, according to the sensitivity analysis performed for the needs of this research work related to the beam deflection.

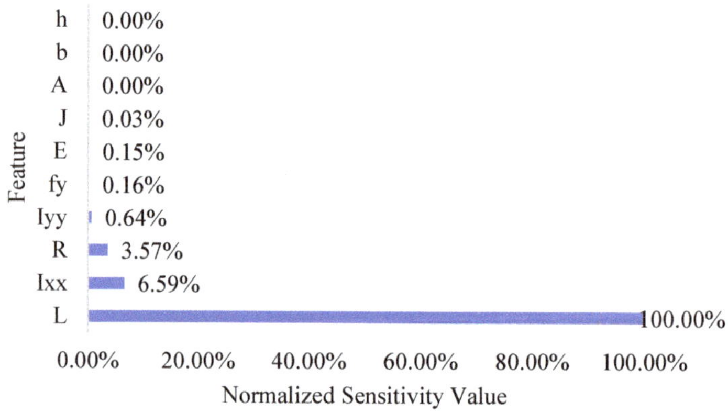

Figure 22. Summary of sensitivity analysis findings (XGBoost-HYT-CV).

It is important to show here that the predictive models did not overfit and that the solution obtained through the ML analysis led to an objective predictive model that did not overfit. In addition to the validation presented in Section 8, Figure 23 shows the tuned cross-validation histories for the DANN-MPIH-HYT and the XGBoost-HYT-CV ML algorithms. It was evident that the algorithms were able to derive models that were optimised through the training and testing procedure.

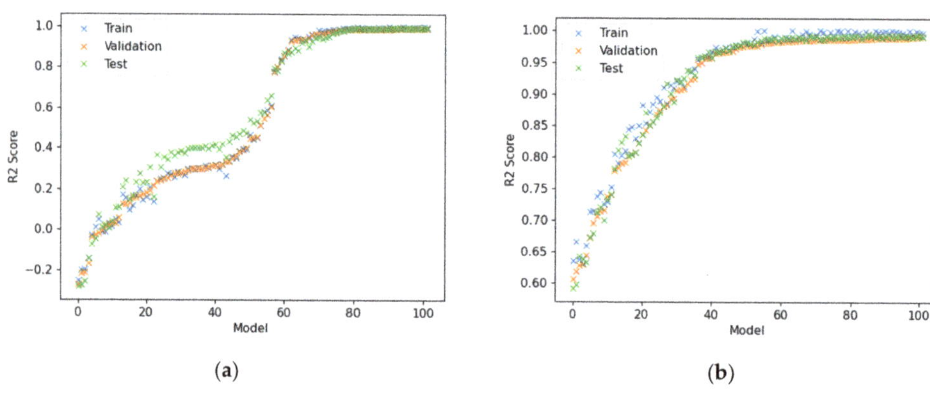

Figure 23. Case of deflection. Tuned cross-validation history for the case of (**a**) DANN-MPIH-HYT and (**b**) XGBoost-HYT-CV.

7.2. Proposed Predictive Models for the Case of Failure Load

This section outlines the failure load predictions provided by the various ML algorithms. An analysis was conducted to determine the performance of the various proposed predictive models. The correlation for the train and test datasets can be seen in Figures 24–27.

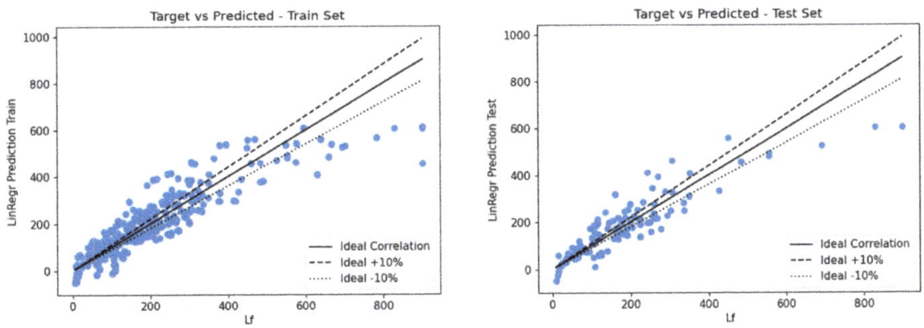

(a) (b)

Figure 24. Correlation between the failure load determined using FEM and the failure load estimated using LR (**a**) on the training dataset (**b**) on the testing dataset.

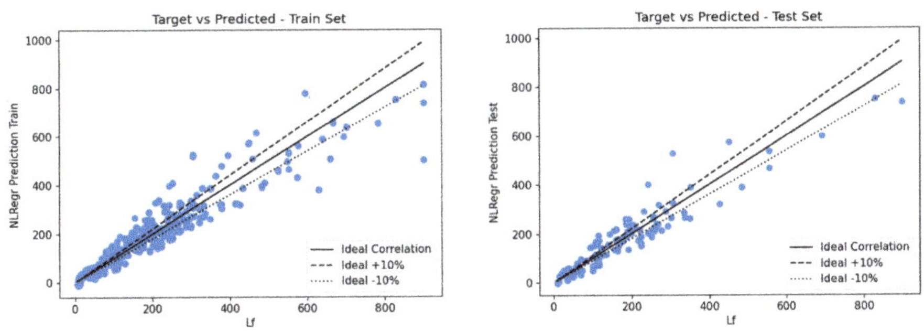

(a) (b)

Figure 25. Correlation between the failure load determined using FEM and the failure load estimated using POLYREG-HYT (**a**) on the training dataset (**b**) on the testing dataset.

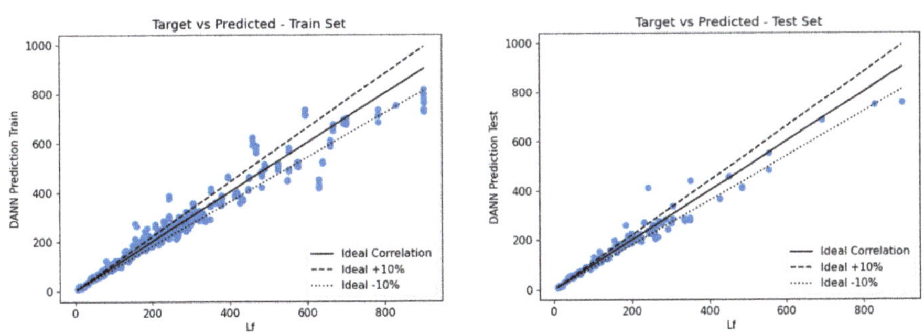

(a) (b)

Figure 26. Correlation between the failure load determined using FEM and the failure load estimated using the DANN-MPIH-HYT algorithm (**a**) on the training dataset (**b**) on the testing dataset.

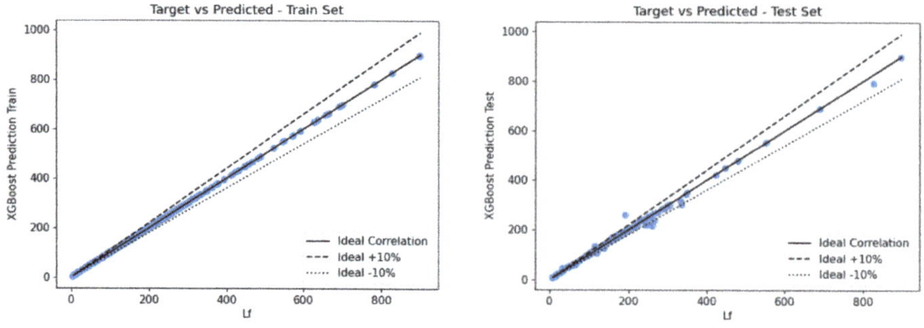

(**a**) (**b**)

Figure 27. Correlation between the failure load determined using FEM and the failure load estimated using the XGBoost-HYT-CV algorithm (**a**) on the training dataset (**b**) on the testing dataset.

As stated previously, it was necessary to use error metrics to determine which ML algorithm performed the best. Therefore, a summary of error metrics can be seen in Table 5. As can be seen, the proposed XGBoost-HYT-CV model was the most accurate when looking at the training and testing datasets. This indicates the fact that no ML algorithm was capable of providing accurate results for all datasets, and different algorithms were used in allocating the best fit to a specific dataset.

Table 5. Comparison of the ML algorithms trained on the failure load dataset.

Model	Dataset	R	MAPE	MAMPE	MAE	RMSE	Computation Time (s)
LR	Train	0.8939	53.64%	27.45%	54.597	76.468	0.001
POLYREG-HYT	Train	0.9369	28.10%	19.26%	38.308	59.633	11.09
DANN-MPIH-HYT	Train	0.9811	10.29%	8.90%	17.693	34.055	2014.47
XGBoost-HYT-CV	Train	1.0000	0.00%	0.00%	0.0025	0.0041	75.07
LR	Test	0.8909	46.06%	27.56%	49.658	67.978	0.00
POLYREG-HYT	Test	0.9412	29.15%	20.86%	37.593	50.776	0.00
DANN-MPIH-HYT	Test	0.9791	10.59%	10.12%	18.241	32.1723	0.0008
XGBoost-HYT-CV	Test	0.9978	3.79%	2.61%	4.694	10.225	0.003

For the cases of the closed-form solutions, Equation (5) shows the formula generated with LR, and Equation (6) provides the proposed predictive formula generated through the POLYREG-HYT ML algorithm:

$$L_f = 2.0577 \times 10^{-6} \times I_{xx} - 1.5497 \times 10^{-4} \times I_{yy} + 7.7002 \times 10^{-4} \times J + 1.9789 \times h - 8.2286 \times b + 2.1078 \times 10^{-1} \times A - 2.0335 \times 10^{1} \times L + 2.5421 \times 10^{-1} \times R + 5.8542 \times 10^{-1} \times f_y - 9.3538 \times 10^{-2} \times E \quad (5)$$

$$
\begin{aligned}
L_f = {} & 1.2526 \times 10^{-3} \times L \times R^2 - 2.153 \times 10^{-4} \times I_{yy} - 1.455 \times 10^{-3} \times b \times A \\
& +1.8898 \times 10^{-5} \times h \times L \times E - 1.2804 \times 10^{-9} \times I_{xx} \times h \times L \\
& -7.8792 \times 10^{-5} \times h \times R^2 + 1.1314 \times 10^{-11} \times I_{xx} \times h \times R \\
& +6.9077 \times 10^{-3} \times L \times f_y - 1.9573 \times 10^{-3} \times R \times f_y \\
& +1.6174 \times 10^{-8} \times I_{yy} \times L^2 + 2.9069 \times 10^{-11} \times I_{yy} \times A \times f_y \\
& +2.2861 \times 10^{-9} \times I_{yy} \times A \times L - 2.2406 \times 10^{-9} \times I_{xx} \times b \times L \\
& +3.1524 \times 10^{-8} \times J \times h \times f_y + 3.8102 \times 10^{-4} \times h \times b^2 \\
& +8.3911 \times 10^{-5} \times b \times R \times E - 2.3274 \times 10^{-6} \times J \times b \\
& -1.1303 \times 10^{-2} \times A \times L - 6.1153 \times 10^{-10} \times I_{yy} \times R \times E \\
& -4.5723 \times 10^{-4} \times b \times L \times f_y - 1.7705 \times 10^{-11} \times J^2 \times f_y \\
& +2.7631 \times 10^{-3} \times f_y^2 - 6.0399 \times 10^{-6} * f_y^3 + 7.0583 \times 10^{-5} \times J \times L \\
& +1.36958 \times 10^{-10} \times I_{xx} \times A - 2.8373 \times 10^{-9} \times J \times R \times E
\end{aligned} \tag{6}
$$

Sensitivity analyses were also conducted on the failure load dataset. A graphical summary of the findings can be seen in Figure 28. As can be seen, the three most influential variables on the midspan failure load were the moment of inertia around the major axis (I_{xx}), the curved length (L), and the radius of curvature (R). I_{xx} was by far the most influential variable, due to the fact that as the stiffness of the beam in bending increased, a larger load was required to reach failure. The cross-sectional area (A), flange width (b), cross-sectional height (h), polar moment of inertia (J), and moment of inertia around the minor axis (I_{yy}) had practically no impact on the failure load according to the XGBoost-HYT-CV ML algorithm. These findings can be used to create an improved, simpler dataset in the future.

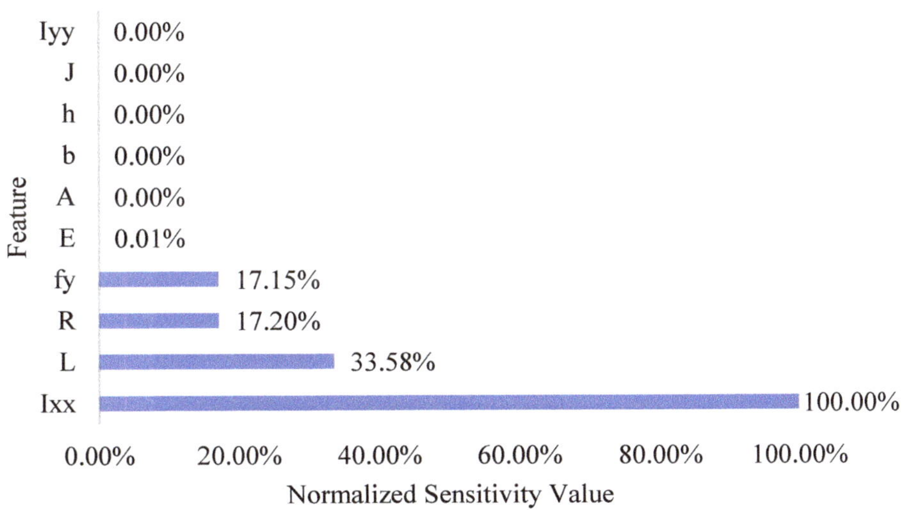

Figure 28. Results of sensitivity analysis (XGBoost-HYT-CV).

Before moving to the validation section presented next, Figure 29 shows the tuned cross-validation history for the case of the DANN-MPIH-HYT and XGBoost-HYT-CV algorithms resulting from the analysis. Once more, it is easy to observe the numerical response of the two histories that converge to R^2 = 1.0, which represents maximum data correlation.

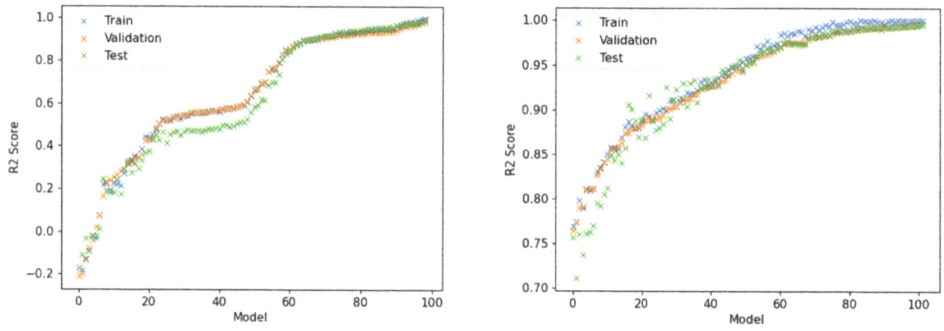

(a) (b)

Figure 29. Case of ultimate load. Tuned cross-validation history for the case of (**a**) DANN-MPIH-HYT and (**b**) XGBoost-HYT-CV.

8. Validation

This section outlines the process of validation that was performed through the use of out-of-sample FE models. The validation aimed to evaluate the proposed predictive models' abilities to capture data that were not used during training or testing. It must be noted here that the ML algorithms were trained on IPE sections (IPE 100, IPE 200, IPE 300, IPE 400, IPE 500, and IPE 600) of various geometries. For validation purposes, it was decided to use sections as per the *South African Steel Construction Handbook* (SASCH). The sections considered were 203 × 133 × 25, 305 × 165 × 40, and 457 × 191 × 67. Therefore, the proposed predictive models had never been exposed to this type of sectional geometry. Additionally, out-of-sample R/L ratios of 2.5 and 5 were considered and the overall span of the section varied depending on the section depth. Out-of-sample yield strengths of 285 MPa and 325 MPa were considered and the Young's modulus values considered were 195 GPa and 205 GPa. This led to a total of 48 out-of-sample beams being created for validation purposes. The descriptive statistics relating to the new validation dataset can be seen in Table 6.

Table 6. Validation models descriptive statistics.

	Mean	Median	Std	Min	Max	Skewness	Kurtosis
I_{xx} (mm^4)	134,333,333	85,500,000	116,928,298	23,500,000	294,000,000	0.58	−1.53
I_{yy} (mm^4)	8,416,667	7,660,000	4,738,358	3,090,000	14,500,000	0.25	−1.53
J (mm^4)	194,667	149,000	134,794	59,000	376,000	0.49	−1.53
h (mm)	320.20	303.8	103.97	203.2	453.60	0.24	−1.53
b (mm)	162.8	165.1	23.37	133.4	189.90	−0.15	−1.53
A (mm^2)	5643.33	516	2225.95	3220	8550	0.33	−1.53
L (m)	9.17	8.5	5	3	18	0.56	−0.75
R (m)	34.38	27.5	22.9	7.5	90	1.11	0.69
f_y (MPa)	305	305	20.21	285	325	0.00	−2.09
E (GPa)	200	200	5.05	195	205	0.00	−2.09

8.1. Validation of Deflection ML Results

The proposed predictive models that were developed to estimate the deflection of the beams were used to predict the deflections of the out-of-sample data. This section focuses only on the two algorithms, namely, DANN-MPIH-HYT and XGBoost-HYT-CV, that were found to outperform the rest of the ML-generated predictive models. Correlation plots of the two ML models can be seen in Figures 30 and 31. These results were compared with the current analytical method, Castigliano's second theorem. The graph showcasing the correlation of Castigliano's second theorem can be seen in Figure 32.

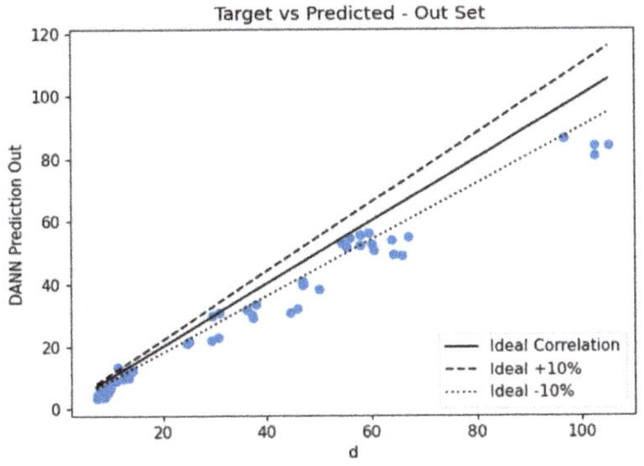

Figure 30. Correlation between the deflection determined using FEM and the deflection estimated using DANN-MPIH-HYT on the validation dataset.

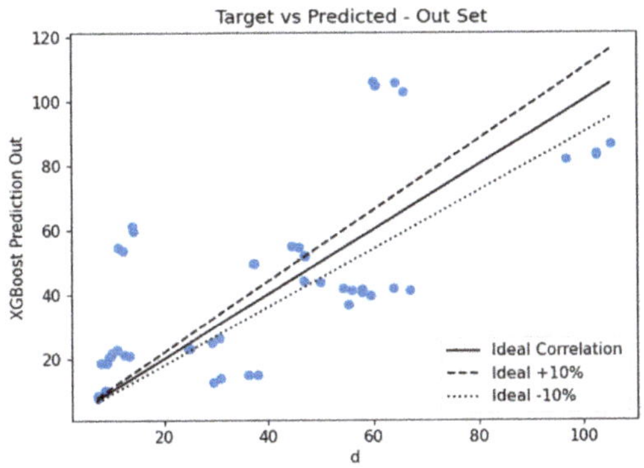

Figure 31. Correlation between the deflection determined using FEM and the deflection estimated using XGBoost-HYT-CV on the validation dataset.

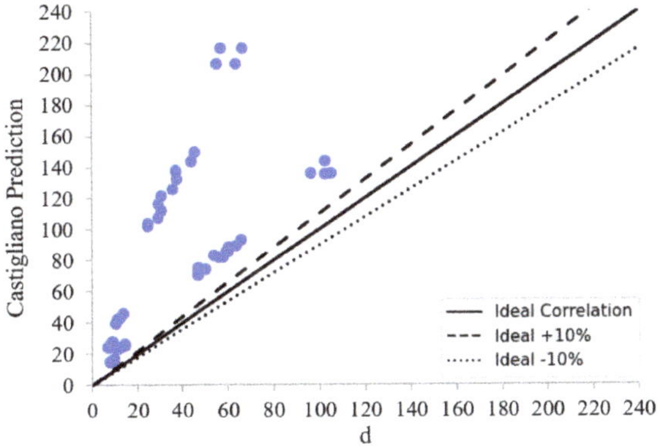

Figure 32. Correlation between the deflection determined using FEM and the deflection estimated using Castigliano's theorem on the validation dataset.

A summary of the error metrics of all ML algorithms can be seen in Table 7. As can be seen, the DANN-MPIH-HYT algorithm outperformed all ML algorithms and the currently used analytical formula. The POLYREG-HYT formula performed the worst. This poor performance was attributed to over-fitting, as the LR formula was quite accurate. It should be noted that even though these results were verified using FEM models, the FE modelling technique used was experimentally verified using an experimental beam where the error experienced was negligible.

Table 7. Summary of the error metrics on the validation dataset (case of deflection).

Algorithm	R	MAPE	MAMPE	MAE	RMSE
LR	0.9267	82.37%	51.84%	20.471	24.426
POLYREG-HYT	0.5174	122.73%	131.81%	52.047	72.419
DANN-MPIH-HYT	0.988	19.08%	16.19%	6.393	8.389
XGBoost-HYT-CV	0.7007	68.37%	42.40%	16.744	21.509
Castigliano	0.6781	155.42%	126.23%	49.842	65.574

8.2. Validation of Failure Load ML Results

The proposed predictive models previously discussed were used to predict the failure load of the out-of-sample data. This section outlines the findings. No analytical formulae are discussed in this section, as the focus of this article is more on deflection estimation and not failure mode investigation. Due to the complexities associated with the failure modes of horizontally curved beams, numerous equations were consulted, and their inclusion would overwhelm the article. The focus herein is on DANN-MPIH-HYT and XGBoost-HYT-CV since they were found to outperform the other predictive models. Correlation plots can be seen in Figures 33 and 34.

The results paint a different picture from what was seen during training and testing. The XGBoost-HYT-CV did not seem to achieve better results than those generated with the DANN-MPIH-HYT algorithm; however, a closer look at the error metrics was required. A summary of the error metrics can be seen in Table 8.

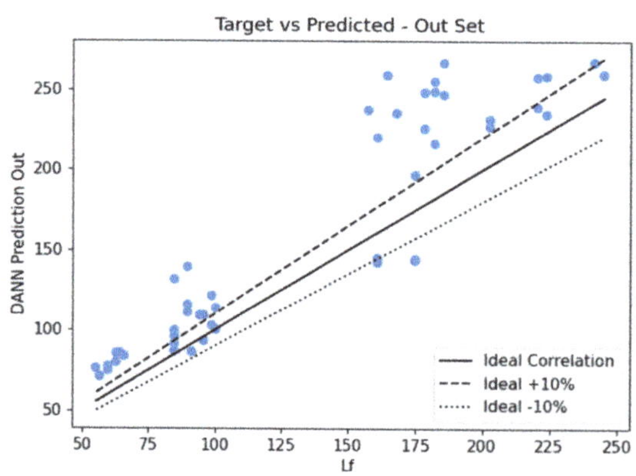

Figure 33. Correlation between the failure load determined using FEM and the failure load estimated using DANN-MPIH-HYT on the validation dataset.

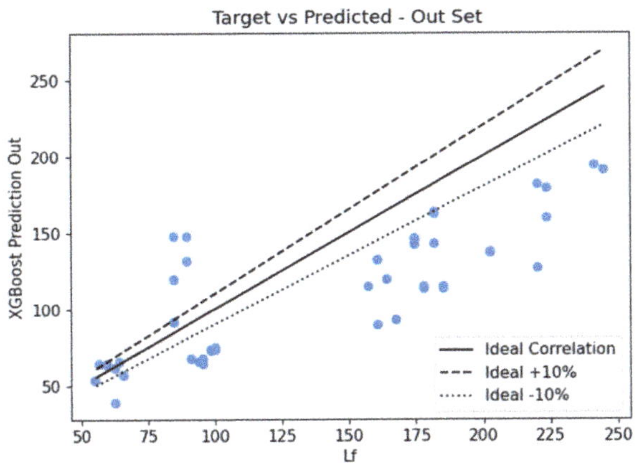

Figure 34. Correlation between the failure load determined using FEM and the failure load estimated using XGBoost-HYT-CV on the validation dataset.

Table 8. Summary of error metrics on the validation dataset (case of failure load).

Algorithm	R	MAPE	MAMPE	MAE	RMSE
LR	0.1440	478.86%	403.82%	548.0527	560.4963
POLYREG-HYT	0.2843	539.71%	500.12%	678.7503	885.5885
DANN-MPIH-HYT	0.9235	23.22%	22.42%	30.4302	38.4831
XGBoost-HYT-CV	0.7914	26.42%	26.87%	36.4645	42.9955

As can be seen, the correlation was significantly worse and all the error metrics showed that the predictions provided through the XGBoost-HYT-CV were less accurate than the predictions provided through the DANN-MPIH-HYT algorithm. This is the same phenomenon as was experienced in the deflection dataset where the DANN-MPIH-HYT algorithm performed better with the out-of-sample data compared with all the other ML algorithms. Therefore, the extended duration required to train the DANN-MPIH-HYT algorithm can be said to translate to improved accuracy during validation.

9. Conclusions and Recommendations

This study showed that FE modelling was able to replicate the experimental results acquired with horizontally curved steel I-beams. For the needs of this research work, an experiment was performed that involved the loading of an IPE100 curved steel I-beam under a vertical load. The experimental data obtained were then used to validate numerical models through the use of CivilFEM software. The usage of quadratic solid FEs with a hexahedral mesh size of 50 mm as well as appropriate material constitutive models paired with nonlinear analyses led to the most accurate results when compared with the experimental data.

A parametric investigation was conducted in which it was noted that the FE modelling technique outlined in this research work was capable of accurately estimating the midspan deflection and rotation of horizontally curved steel I-beams. The MAE was noted to be 3.32% when estimating deflection and 9.17% when estimating rotation, which far outperformed analytical methods, which had an error of 19.80% when estimating deflection and 16.07% when estimating rotation.

The experimentally validated model was then used to develop numerous FE meshes that were analysed under ultimate limit state loading conditions. The numerically obtained results were used to develop a large dataset. A total of 864 models were created that encompassed the entire IPE cross-section list (IPE100 to IPE600). A total of 10 independent variables were considered in this study.

Numerous ML algorithms were used, namely, LR, POLYREG-HYT, DANN-MPIH-HYT, and XGBoost-HYT-CV, for developing the proposed predictive models. During the validation phase, experimentally validated FE models that were outside the training dataset were created to validate the various proposed predictive models proposed in this research work. To further evaluate the accuracy of the available analytical methods for computing the deflection and ultimate load of curved steel I-beams, the validation data were used to assess Castilgiano's analytical formulae as well. According to the numerical findings, the DANN-MPIH-HYT algorithm was the most accurate in estimating both deflection and failure load. When estimating deflection, the DANN-MPIH-HYT proposed predictive model had a MAMPE of 16.19%. This was found to be a significant improvement compared with the analytical method, Castigliano's second theorem, which derived an extremely large MAMPE of 126.23%, highlighting the need for more accurate and objective predictive models. It is also safe to conclude that based on the findings of this dissertation, the proposed ML-generated predictive models are far more accurate than the current analytical methods for estimating the deflection of curved steel I-beams.

In addition to the above, when estimating the failure load of curved steel I-beams, the DANN-MPIH-HYT generated model was found to be significantly more accurate than the other proposed models. This is most likely to be due to the complexities associated with horizontally curved steel I-beams and the nature of the datasets. Horizontally curved steel I-beams have various failure modes, given that they can fail due to flexure, torsion, shear, or a combination of these. The beams may also fail due to lateral torsional buckling. Therefore, a larger dataset is required for the ML algorithms to derive the patterns that are connected to the failure modes of the beams. At this stage, the international literature includes numerous equations for each of the failure modes. Therefore, there is no individual formula to compare the estimates with. It is believed that the DANN-MPIH-HYT algorithm, which resulted in a MAMPE of 22.42%, can be further improved in the future by either increasing the dataset or providing different formulae for the different failure modes, as is currently the case in all design codes.

Finally, this research study is able to propose, for the first time in the international literature, accurate and objective predictive models that outperform any known formula used to compute the deflection of curved steel I-beams and their ultimate capacity. The largest datasets currently available in the international literature were also developed for the training and testing of the proposed predictive models. This pilot project paves the way for the development of future design formulae that can be more accurately applied to a larger range of beams for different boundary conditions.

This research was limited to evaluating the feasibility of using FE models to estimate the midspan deflection and failure load of horizontally curved steel I-beams with a single point load at midspan and fully fixed at both ends. The development of the datasets foresaw the use of minimum and maximum geometrical features. Therefore, the proposed predictive models should not be used for beams with dimensions that are larger or smaller than the respective maximum and minimum geometrical values of the beams found in the datasets developed herein. Furthermore, recommendations for future work should include the following:

- Perform more experiments on curved steel I-beams to further validate the proposed predictive models;
- Consider various boundary conditions, such as torsionally pinned beams, and determine the influence this has on deflection, failure load, rotation, and stress distribution. In theory, this should not significantly impact the failure load. However, this is expected to have a drastic impact on the stress distribution and rotation of the section;
- Consider various loading conditions, such as uniformly distributed loads;
- Consider the impact residual stresses have on stress distribution and investigate whether this has an impact on deflection and failure loads. Residual stresses are far

greater in horizontally curved steel I-beams compared with straight beams, due to the initial cold-forming process. Therefore, residual stresses vary depending on the R/L ratio at hand. Currently, there is no formula available in the international literature that allows accounting for how the residual stresses vary throughout the length of the beam and within the cross-section. Therefore, a detailed investigation is required.
- Develop an accurate formula to determine the midspan rotation of horizontally curved steel I-beams using a relevant dataset.

Author Contributions: Conceptualization, G.M. and E.A.; Methodology, G.M., E.A. and S.S.; Software, E.A. and G.M.; Validation, E.A. and G.M.; Formal Analysis, E.A.; Investigation, E.A.; Resources, G.M. and E.A.; Data curation, E.A.; Writing—original draft preparation, E.A.; Writing—Reviewing and editing, G.M. and S.S.; Visualization, E.A. and G.M.; Supervision, G.M. and S.S.; Project Administration, G.M.; Funding Acquisition, E.A. and G.M. All authors have read and agreed to the published version of the manuscript.

Funding: This research was funded by the National Research Fund (NRF South Africa [MND21062361 5086]) and the APC was funded by the journal.

Data Availability Statement: The datasets that were developed for the needs of this research work can be found through the following link (https://github.com/nbakas/nbml/tree/a0d27c94dd59068 8815180ebf6428963a24ca245/datasets, accessed on 1 July 2024).

Conflicts of Interest: The authors declare no conflict of interest.

References

1. Liew, J.R.; Thevendran, V.; Shanmugam, N.; Tan, L. Behaviour and design of horizontally curved steel beams. *J. Constr. Steel Res.* **1995**, *32*, 37–67. [CrossRef]
2. Sonck, D.; Van Impe, R. Study of residual stresses in I-section members and cellular members. In Proceedings of the Annual Stability Conference Structural Stability Research, St. Louis, MO, USA, 16–20 April 2013.
3. Rossini, N.; Dassisti, M.; Benyounis, K.; Olabi, A.-G. Methods of measuring residual stresses in components. *Mater. Des.* **2012**, *35*, 572–588. [CrossRef]
4. King, C.; Brown, D. *Design of Curved Steel*; Steel Construction Institute Ascot: Ascot, UK, 2001.
5. Rettie, N.J. Lateral Torsional Buckling Resistance of Horizontally Curved Steel I-Girders. MSc Thesis, University of Alberta, Edmonton, AB, Canada, 2015.
6. Dowsell, B. *Design Guide 33: Curved Member Design*; American Institute of Steel Construction: Chicago, IL, USA, 2018.
7. Lee, K.; Davidson, J.S.; Choi, J.; Kang, Y. Ultimate strength of horizontally curved steel I-girders with equal end moments. *Eng. Struct.* **2017**, *153*, 17–31. [CrossRef]
8. Wong, Y.-C. Horizontally Curved Beam Analysis and Design. MSc Thesis, Oregon State University, Corvallis, OR, USA, 1969.
9. Dahlberg, T. Procedure to calculate deflections of curved beams. *Int. J. Eng. Educ.* **2004**, *20*, 503–513.
10. Shanmugam, N.; Thevendran, V.; Liew, J.R.; Tan, L. Experimental study on steel beams curved in plan. *J. Struct. Eng.* **1995**, *121*, 249–259. [CrossRef]
11. Shanmugam, N.; Mahendrakumar, M.; Thevendran, V. Ultimate load behaviour of horizontally curved plate girders. *J. Constr. Steel Res.* **2003**, *59*, 509–529. [CrossRef]
12. Dabaon, M.; El-Naggar, M.; Yossef, N. Experimental and theoretical study of curved rolled and castellated composite beams. *Alex. Eng. J.* **2003**, *42*, 219–230.
13. Lim, J.; Kang, Y.-J.; Lee, J.; Kim, S.; Lee, K. Experimental and Analytical Study of Horizontally Curved I-Girders Subjected to Equal End Moments. *Metals* **2021**, *11*, 1132. [CrossRef]
14. Markou, G.; Bakas, N.; Papadrakakis, M.; Chatzichristofis, S. A General Framework of High-Performance Machine Learning Algorithms: Application in Structural Mechanics. *Comput. Mech.* **2024**, *73*, 705–729. [CrossRef]
15. Gravett, D.Z.; Mourlas, C.; Taljaard, V.-L.; Bakas, N.; Markou, G.; Papadrakakis, M. New fundamental period formulae for soil-reinforced concrete structures interaction using machine learning algorithms and ANNs. *Soil Dyn. Earthq. Eng.* **2021**, *144*, 106656. [CrossRef]
16. Markou, G.; Bakas, N.P. Prediction of the shear capacity of reinforced concrete slender beams without stirrups by applying artificial intelligence algorithms in a big database of beams generated by 3D nonlinear finite element analysis. *Comput. Concr.* **2021**, *28*, 433–447.
17. AlHamaydeh, M.; Markou, G.; Bakas, N.; Papadrakakis, M. AI-based shear capacity of FRP-reinforced concrete deep beams without stirrups. *Eng. Struct.* **2022**, *264*, 114441. [CrossRef]
18. Calitz, D.; Markou, G.; Bakas, N.; Papadrakakis, M. Developing Fundamental Period Formulae for Steel Framed Structures through Machine Learning and Automated Algorithms. In Proceedings of the Compdyn 2023, Athens, Greece, 12–14 June 2023.

19. Chibaya, V.T.; Markou, G.; Bakas, N. Development of Formulae for the Section Rotations Due to Bending of Curved Steel I-Beams through Ai and Ml Algorithms. In Proceedings of the Compdyn 2023, Athens, Greece, 12–14 June 2023.
20. van der Westhuizen, A.M.; Bakas, N.; Markou, G. Developing an Artificial Neural Network Model That Predicts the Fundamental Period of Steel Structures Using a Large Dataset. In Proceedings of the Compdyn 2023, Athens, Greece, 12–14 June 2023.
21. Ababu, E.; Markou, G.; Bakas, N. Using Machine Learning and Finite Element Modelling to Develop a Formula to Determine the Deflection of Horizontally Curved Steel I-Beams. In Proceedings of the 14th International Conference on Agents and Artificial Intelligence, Virtual, 3–5 February 2022; pp. 958–963. [CrossRef]
22. Aminu, I.M.; Shariff, M.N.M. Strategic orientation, access to finance, business environment and SMEs performance in Nigeria: Data screening and preliminary analysis. *Eur. J. Bus. Manag.* **2014**, *6*, 124–132.
23. Dimopoulos, T.; Tyralis, H.; Bakas, N.P.; Hadjimitsis, D. Accuracy measurement of Random Forests and Linear Regression for mass appraisal models that estimate the prices of residential apartments in Nicosia, Cyprus. *Adv. Geosci.* **2018**, *45*, 377–382. [CrossRef]

computation

MDPI

Article

Explainable Boosting Machine Learning for Predicting Bond Strength of FRP Rebars in Ultra High-Performance Concrete

Alireza Mahmoudian [1], Maryam Bypour [2] and Mahdi Kioumarsi [3,*]

[1] Department of Civil Engineering, Shahid Rajaee Teacher Training University, Tehran P.O. Box 16788-15811, Iran; amahmoudian1997@sru.ac.ir
[2] Department of Civil Engineering, Semnan University, Semnan 1581613711, Iran; mbypour@gmail.com
[3] Department of Built Environment, OsloMet—Oslo Metropolitan University, 0166 Oslo, Norway
* Correspondence: mahdi.kioumarsi@oslomet.no; Tel.: +47-67-23-87-45

Abstract: Aiming at evaluating the bond strength of fiber-reinforced polymer (FRP) rebars in ultra-high-performance concrete (UHPC), boosting machine learning (ML) models have been developed using datasets collected from previous experiments. The considered variables in this study are rebar type and diameter, elastic modulus and tensile strength of rebars, concrete compressive strength and cover, embedment length, and test method. The dataset contains two test methods: pullout tests and beam tests. Four types of rebar, including carbon fiber-reinforced polymer (CFRP), glass fiber-reinforced polymer (GFRP), basalt, and steel rebars, were considered. The boosting ML models applied in this study include AdaBoost, CatBoost, Gradient Boosting, XGBoost, and Hist Gradient Boosting. After hyperparameter tuning, these models demonstrated significant improvements in predictive accuracy, with XGBoost achieving the highest R^2 score of 0.95 and the lowest Root Mean Square Error (RMSE) of 2.21. Shapley values analysis revealed that tensile strength, elastic modulus, and embedment length are the most critical factors influencing bond strength. The findings offer valuable insights for applying ML models in predicting bond strength in FRP-reinforced UHPC, providing a practical tool for structural engineering.

Keywords: ultra-high-performance concrete; machine learning; gradient boosting; XGBoost; shapley values technique

1. Introduction

Fiber-reinforced polymer (FRP) rebars have been introduced as an alternative to address the corrosion challenges associated with traditional steel reinforcements [1]. FRP rebars offer corrosion resistance and a high strength-to-weight ratio, making them an attractive choice for concrete structures [2]. Compared to conventional steel rebars, FRP rebars exhibit distinct characteristics, such as high tensile strength and lightweight properties. However, unlike steel rebars, FRP rebars demonstrate no plastic behavior (yielding) before rupture, highlighting their unique tensile behavior [3].

Various types of FRP rebars, including glass FRP (GFRP) [4–7], carbon FRP (CFRP) [8], basalt FRP (BFRP) [9–11], and aramid FRP (AFRP) [12], have emerged as promising alternatives to traditional steel reinforcements. Each type of FRP rebar possesses unique characteristics with advantages and limitations. Generally, FRP rebars are brittle polymers with a lower modulus of elasticity compared to steel, particularly in the case of GFRP and BFRP bars [13].

GFRP rebars are particularly popular among the different FRP types due to their cost-effectiveness. The ratio of FRP reinforcement significantly influences the flexural capacity and failure mode of concrete beams. Increasing the GFRP reinforcement ratio has proven more effective in enhancing beam flexural capacity than adding steel fibers or optimizing fiber orientations [14]. Despite their appeal, GFRP rebars face challenges such as limited toughness, weaker bonding with concrete compared to steel rebars, and low fire resistance [15].

Although current FRP design standards and specifications do not extensively incorporate BFRP bars due to limited studies on their durability, basalt fiber offers promising attributes, including the ability to withstand high temperatures. Basalt fiber is an environmentally friendly material classified as sustainable due to its natural composition and the absence of chemical additives during production. It is considered a "green" material derived from rock [16].

One of the significant concerns with using CFRP in reinforced concrete (RC) members is the low bonding property of CFRP rods embedded in concrete, mortar, and epoxy resin. To address this issue, research has investigated using CFRP rods with attached GFRP ribs to improve bond strength. The study demonstrated excellent load-carrying capacity and fatigue durability of RC members strengthened by CFRP rods with ultra-high modulus [17].

Recently, ultra-high-performance fiber-reinforced concrete (UHPFRC) has gained significant research attention [18–20]. UHPFRC is a cement-based material characterized by high compressive strength, tensile and flexural strengths, ductility, and remarkable durability [21]. Ultra-high-performance concrete (UHPC) beams reinforced with FRP bars have exhibited high flexural stiffness and minimal crack width at the serviceability limit state [22].

Research on the bond performance of CFRP bars in UHPFRC, including tests on pullout specimens, revealed that bond strength increases with larger CFRP bar diameters. A theoretical model was developed to predict bond strength [23]. Additional studies have shown that the bond performance between CFRP bars and UHPC is predominantly affected by pullout damage, with CFRP bar surfaces peeling off from the internal core while UHPC remains undamaged. Enhancing the cover and steel fiber volume fraction improves bonding performance, whereas increasing the bar diameter reduces it. Equations for calculating ultimate bond strength and development length have been proposed, integrating factors such as CFRP bar diameter, bonded length, and cover thickness [24]. Additionally, the combined use of CFRP and UHPC has demonstrated strong performance as a retrofitting method for pre-damaged concrete [25].

The performance of concrete elements reinforced with FRP bars is fundamentally influenced by the bond properties between the reinforcement and the surrounding concrete. Achieving a sufficient level of bonding is critical for ensuring effective force transmission between these two materials. The substitution of steel with FRP significantly alters the load transfer mechanism between the concrete and reinforcement. The tensile behavior of FRP bars, which are composed of a single type of fiber material, is characterized by a linear elastic stress-strain response up to the point of failure [3].

While several practical formulas for calculating the ultimate bond strength of FRP bars in concrete exist in standards such as ACI 440.1R-06 [3] and CSA S806-12 [26], these codes primarily target ordinary concrete. They may not directly apply to evaluate the bond performance between FRP bars and UHPC with high compressive strength [27].

Recent advancements in machine learning (ML), particularly in boosting algorithms, have shown great promise in predicting complex material behavior in structural engineering. Boosting techniques such as AdaBoost, Gradient Boosting Machine, and XGBoost have been successfully applied to predict properties such as the compressive strength of ordinary concrete and high-performance concrete (HPC) [28–32]. The effectiveness of AdaBoost in predicting concrete compressive strength with high accuracy has been proven, outperforming the other ML methods, such as artificial neural networks [30]. Similarly, the Gradient Boosting Machine method was employed to model the nonlinear relationships in high-performance concrete [31], and XGBoost was used to predict CNT-modified concrete's compressive strength [32]. These studies highlight the ability of boosting algorithms to handle complex datasets and deliver accurate predictions, making them suitable for analyzing material properties in structural engineering contexts.

Numerous studies have investigated the bond strength of various FRP rebars in UHPFRC using different testing methods, such as pullout and beam tests [33–38]. This study developed boosting ML-based models to predict the bond strength of UHPFRC containing various FRP bar types and test methods. Due to the limited dataset size typical in civil engineering research, non-parametric ML models are more suitable as they can effectively handle smaller datasets without overfitting. These models were chosen for their robustness, interpretability, and superior performance in capturing complex interactions within small datasets. The research also examined the significance of multiple features on the bond strength of UHPFRC. Variables considered in this study include rebar type and diameter, elastic modulus and tensile strength of rebars, concrete compressive strength, embedment length, and test method. The dataset includes two test methods—pullout and beam tests—and four types of rebars, including CFRP, GFRP, basalt, and steel rebars. Figure 1 illustrates the overall process of the study, starting with data collection from experimental studies on the bond strength of FRP rebars in UHPC. The data collection process includes various features related to concrete and rebars. These variables were then used as input features for the ML models, including AdaBoost, CatBoost, Gradient Boosting, XGBoost, and Hist Gradient Boosting. The figure highlights how these models were trained and tuned, with the final step showcasing model evaluation using R^2, RMSE, and MAE metrics. Additionally, feature importance analysis was conducted to identify the most influential variables impacting bond strength predictions.

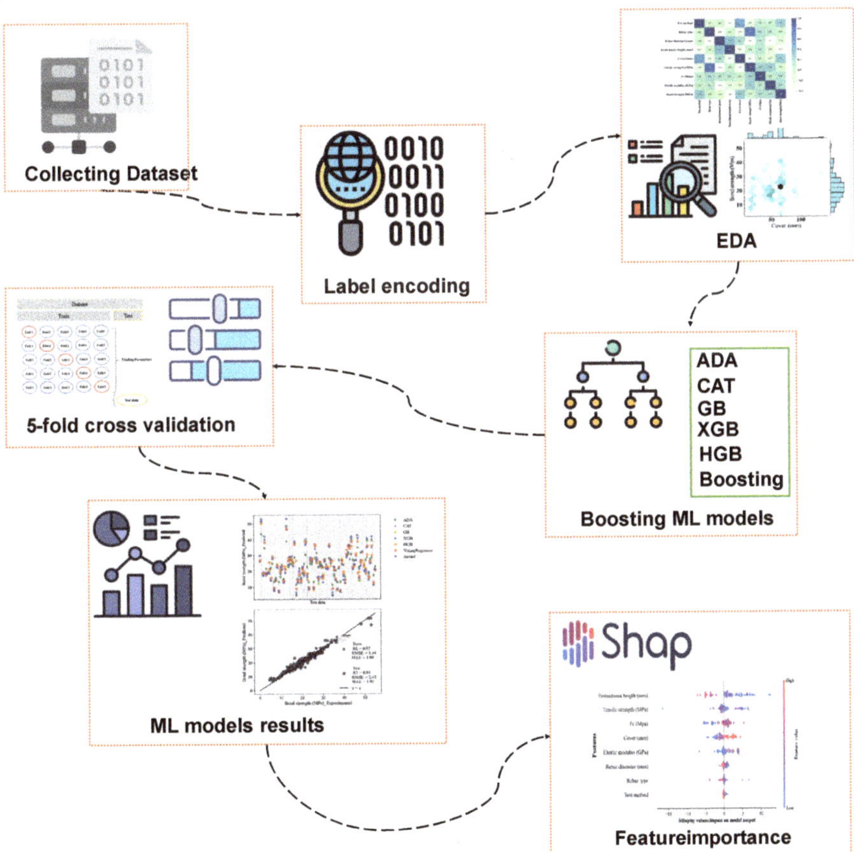

Figure 1. Overview of the study.

2. Previous Experimental Works on Bond Strength Assessment of Various FRP Bars in UHPFRC

Figure 2 depicts the condition of an FRP bar embedded in concrete. An average bond stress resists the stress in the bar, u. The equilibrium of forces can be calculated according to Equation (1).

$$l_e \pi d_b u = A_{fbar} f_f \quad (1)$$

where l_e, d_b, and A_{fbar} represent the embedment length, diameter, and cross-sectional area of the bar, respectively, while f_f denotes the stress developed in the bar at the end of the embedment length. Unlike steel bars, it is not always necessary to fully develop the strength of an FRP bar, particularly in cases where the flexural capacity is controlled by concrete crushing. In such scenarios, the required stress in the FRP bar at the point of failure may be lower than its guaranteed ultimate strength.

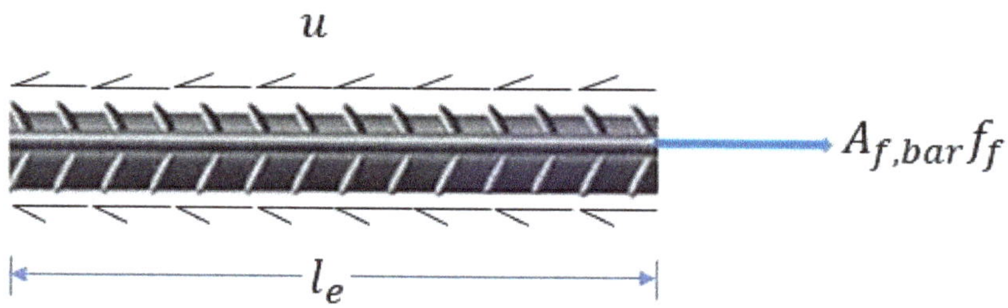

Figure 2. Schematic of FRP bar under uniaxial loading.

Table 1 summarizes previous studies investigating the bond strength of different FRP bars embedded in UHPFRC. Some studies focus on the behavior of individual FRP bar types, such as GFRP, CFRP, and BFRP, alongside conventional steel rebar used as a control specimen. One study, in particular, examined three types of FRP rebar—ribbed CFRP, sand-coated CFRP, and ribbed GFRP—along with steel rebar as a comparative benchmark [39]. The most frequently studied variables in the reviewed literature include rebar type and diameter, embedment length, and concrete cover.

Several findings from these studies were consistent across different FRP rebar types. The results generally indicated that the bond strength between FRP rebars, irrespective of the FRP type, was lower than that of steel rebars [40,41]. However, UHPC beams reinforced with GFRP bars showed a significant increase in flexural capacity compared to steel ones [27]. Furthermore, these findings suggest that using bars with smaller diameters enhances bond capacity [42].

Moreover, it was observed that increasing the embedment length could lead to a reduction in bond strength between FRP bars and UHPC [39,40,42]. However, one study reported an inverse relationship, where the bond strength of helically ribbed CFRP bars in UHPC decreased with a reduction in embedment length [8].

The results also confirmed the positive impact of concrete cover on bond strength, with a continuous increase in bond strength observed as the concrete cover thickness increased [8,40]. A comparison of maximum bond strength revealed the following order: ribbed CFRP bars, ribbed GFRP bars, steel bars, and sand-coated CFRP bars [39].

Additionally, a comparison of test methods indicated that the bond strength measured using the hinged beam test was lower than that obtained from the direct pullout test [33]. Specifically, the bond strength of GFRP rebars was higher in the beam test compared to the pullout test [40].

Table 1. Summary of previous studies investigating the bond strength of various FRP bars in UHPFRC.

Research	Test Type	Fiber Type	Variables	Findings
Hu et al. (2024) [40]	Pullout, beam	GFRP	Embedment length Concrete cover Rebar type	Two distinct bond stress-slip relationships were identified based on the embedment length of GFRP rebars. The bond strength of GFRP rebars was higher in the beam test compared to the pullout test, with the modified pullout test showing only a slight difference. GFRP bars exhibited lower bond strength with UHPC than steel bars, regardless of the testing method. Increasing the embedment length and decreasing the cover led to a linear reduction in bond strength between GFRP bars and UHPC. It is recommended that the development length for sand-coated or deformed GFRP rebars with smaller diameters in UHPC should be at least 13 times the bar diameter, with a cover thickness not less than twice the bar diameter.
Yoo et al. (2023) [8]	Pullout	CFRP	Rebar profile Embedment length Rebar diameter	The critical concrete cover thickness for helically ribbed CFRP bars to prevent splitting failure is greater in high-strength concrete than in normal-strength concrete. Bond strength in UHPC increases consistently with greater concrete cover thickness. As the compressive strength of concrete increases, both the bond strength and bond stiffness of ribbed CFRP bars also improve. Helically, ribbed CFRP bars in UHPC with longer embedment lengths exhibit higher bond strength than shorter lengths. The bond strength of helically ribbed CFRP bars is more than double that of sand-coated CFRP bars. Helically, ribbed CFRP bars demonstrate greater initial and post-toughness than sand-coated bars, although the post-toughness difference narrows due to the friction in sand-coated bars. Existing bond design codes and proposed formulas inadequately predict the bond strength of CFRP bars in UHPC, particularly due to variations in CFRP bar profiles. Bond strength predictions using modified equations and an artificial neural network (ANN) method proved more accurate, with the ANN demonstrating superior predictive capability. A modified bond equation for helically ribbed and sand-coated CFRP bars enhanced prediction accuracy using the ANN approach.
Mahaini et al. (2023) [27]	Four-point loading	GFRP	Reinforcement ratio Number of rebars Surface texture bars	GFRP-UHPC beams exhibited a typical bilinear response in both deflections and strains. All GFRP beams showed similar stiffness during pre-cracking, independent of the reinforcement ratio. Post-cracking stiffness increased with higher GFRP reinforcement ratios. Higher reinforcement ratios improved the energy absorption capacity of the beams, reducing post-cracking strains in the GFRP bars. Increasing the reinforcement ratio also enhanced the flexural capacity of the GFRP-UHPC beams. When maintaining the same axial stiffness, the number of bars had minimal impact on the flexural behavior of UHPC beams. Shifting the failure mode from GFRP rupture to concrete crushing improved the ductility of the UHPC beams. GFRP-reinforced UHPC beams demonstrated significantly higher flexural capacity than steel-reinforced beams due to the higher tensile strength of GFRP. However, steel-reinforced beams had greater stiffness and lower midspan deflection. The ACI equation provided reasonable predictions for under-reinforced beams but was unconservative for over-reinforced beams, overestimating flexural capacity. The CSA code produced better deflection predictions than the ACI440 equations. However, at ultimate capacity, both ACI440 and CSA specifications were unconservative, particularly for over-reinforced beams, as they failed to account for the increased ductility.

Table 1. *Cont.*

Research	Test Type	Fiber Type	Variables	Findings
Yoo et al. (2024) [39]	Pullout	CFRP Ribbed CFRP Sand coated CFRP Ribbed GFRP Steel rebar	Rebar type Embedment length Fibers (with and without fibers in UHPC) Presence of shear reinforcement	Initially, stiffness was highest for steel bars, followed by sand-coated CFRP, ribbed CFRP, and ribbed GFRP bars. However, after the steel bars yielded, the order shifted to sand-coated CFRP, ribbed CFRP, ribbed GFRP, and steel. Rupture occurred in ribbed CFRP, ribbed GFRP, and steel bars at certain bond lengths, while sand-coated CFRP bars did not rupture even at longer bond lengths. The bond strength of FRP bars decreased as the bonded length increased. The maximum bond strength followed the order: ribbed CFRP bars, ribbed GFRP bars, steel bars, and sand-coated CFRP bars. Combining fiber mixing with a reinforcement cage will significantly enhance bond strength and ductility. Due to different stress transfer mechanisms, bond strength measured in the hinged beam test was lower than in the direct pullout test.
Eltantawi et al. (2022) [42]	Four-point loading	BFRP	Rebar diameter Embedment length Rebar surface texture (sand coated (SC) and helically wrapped (HW))	The load-carrying capacities of beams reinforced with SC-BFRP and HW-BFRP bars were nearly identical for the same embedment length. The surface texture of BFRP bars had a minimal effect on the bond with concrete. SC-BFRP bars exhibited slightly higher bond strength compared to HW-BFRP bars. The bond strength of spliced BFRP bars decreased as splice length increased, with longer splice lengths reducing bond strength. Larger diameter bars require longer splice lengths to reach maximum capacity, suggesting that smaller diameter bars enhance splice bond capacity. The ACI 440.1R-15 and CSA S806-12 equations are conservative in predicting splice lengths for BFRP bars, while the CSA-S6-14 equation is more accurate for larger diameters but less so for smaller diameters.
Qasem et al. (2020) [41]	Pullout	CFRP Steel rebars	Rebar type Rebar diameter	Steel rebars exhibit superior bond strength compared to CFRP rebars across various types of concrete. Control specimens without carbon nanotubes (CNTs) showed that steel rebars required higher bond stress for pullout than CFRP rebars due to stronger concrete-steel bonding. Due to their high reactivity, the inclusion of CNT nanoparticles enhances the bond strength between rebars and UHPC. Increasing CNT content in UHPC mix design raises the force needed to pull out steel rebars compared to control specimens. However, excessive CNT content leads to increased porosity due to agglomeration, reducing the bond strength of CFRP rebars in UHPC.

Several studies proposed formulas based on their results and compared them with existing design codes. Bond strength was predicted using two proposed equations with modified bond parameters and an artificial neural network (ANN) method, where the ANN demonstrated superior accuracy over the proposed formulas. A modified bond equation for helically ribbed and sand-coated FRP bars also improved prediction accuracy with the ANN approach [8]. Another study found that the ACI 440 equation provided reasonable predictions for under-reinforced beams but was unconservative for over-reinforced beams, overestimating flexural capacity. The CSA code offered better deflection predictions than ACI 440 equations. However, at ultimate capacity, both ACI 440 and CSA specifications were unconservative, particularly for over-reinforced beams, as neither code accounted for the additional ductility gained by the beams [27]. Lastly, the ACI 440.1R-15 and CSA S806-12 equations were conservative in predicting embedment length for BFRP bars, while the CSA-S6-14 equation was more accurate for BFRP with larger diameters. However, it was not conservative for smaller diameters [42].

3. Dataset Collection

To evaluate the bond strength of various FRP bars in UHPFRC using ML models, a dataset of 249 specimens from existing experimental studies was compiled. The specifics of this dataset are outlined in Table 2. The features analyzed in this study encompass both concrete and rebar characteristics. Table 2 details that the specimens underwent both pullout and beam tests. These test methods are further depicted in Figure 3. ACI 440.3R-12 [43] provides information about these test methods. In the pullout test, the displacement between the free end of the rebar and the UHPC is measured, whereas in the beam test, the displacement is measured at the beam supports [40]. The dataset includes four types of rebar: GFRP, CFRP, basalt, and steel. Rebar diameters range from 7.5 to 20 mm, with embedment lengths varying between 25 and 276 mm. The data reveal that fiber-reinforced rebars exhibit a low modulus of elasticity, ranging from 47 to 158, while all UHPFRC specimens have a high compressive strength (f'_c), between 71 and 181.

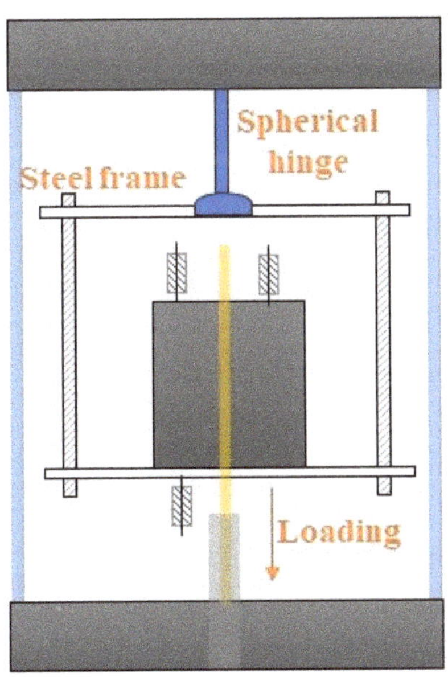

(a)

Figure 3. *Cont.*

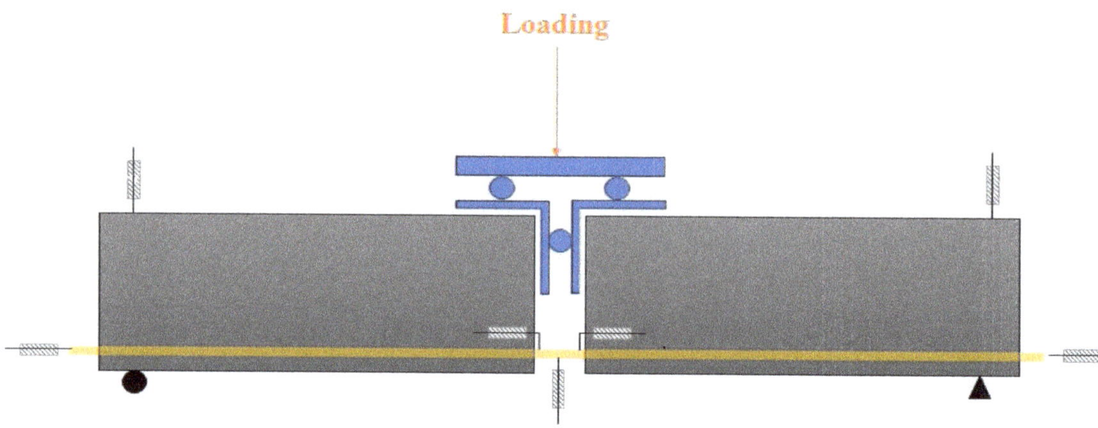

(**b**)

Figure 3. Setup of investigated test methods: (**a**) pullout test, and (**b**) beam test.

Table 2. Experimental dataset collected from existing literature.

Research	Test Method	Rebar Type	Rebar Diameter (mm)	Embedment Length (mm)	Tensile Strength (MPa)	f_c' (MPa)	Elastic Modulus (GPa)	Number of Specimens
Hu et al. (2024) [40]	Pullout and Beam	GFRP and Steel	16	40–160	609 and 894	133.2	54 and 198	16
Zhu et al. (2023) [24]	Beam	CFRP	8, 10, and 12	25–120	2030 and 2702	131–143	-	11
Liang et al. (2023) [13]	Pullout	GFRP, BFRP, and Steel	12	42	355–1321	93–122	48–200.4	48
Tong et al. (2023) [44]	Pullout	GFRP	12, 16, and 20	60–100	702–782	90–132	54–58	54
Decebal et al. (2021) [36]	Beam	GFRP	17.2	69–276	1100	87–132	60	28
Hossain et al. (2017) [33]	Beam	GFRP	15.9 and 19.1	47–133	751–1439	71–174	47–64	48
Ahmed and Sennah (2014) [45]	Pullout	GFRP	20	80–160	1105	166–181	64.7	35
Ahmad et al. (2011) [34]	Pullout	CFRP	7.5, 8, 10, and 12	40–160	2300 and 2400	170	130 and 158	9

CFRP: carbon fiber-reinforced polymer; GFRP: glass fiber-reinforced polymer; BFRP: basalt fiber-reinforced polymer.

4. Dataset Construction

Figure 4 presents the results of the correlation analysis and distribution of key input variables concerning bond strength. Each hexagon represents a data point, and the color gradient in the hexagons indicates the density of the data points in specific regions of the plot. Darker hexagon colors represent regions of higher density, where more data points exhibit similar parameter values, thus signifying stronger correlations between these values and bond strength. Lighter colors indicate areas with fewer data points and weaker correlations. The color intensity helps visualize the concentration of the data and highlights where certain parameter values have more influence on bond strength. In Figure 4, the variables "Test method" and "Rebar type" are represented numerically for visualization purposes. The numeric equivalents for the test methods are as follows: 0 corresponds to the pullout test, while 1 represents the beam test. For the rebar types, the numeric representation is as follows: 0 denotes CFRP, 1 indicates GFRP, 2 stands for BFRP, and 3 signifies steel.

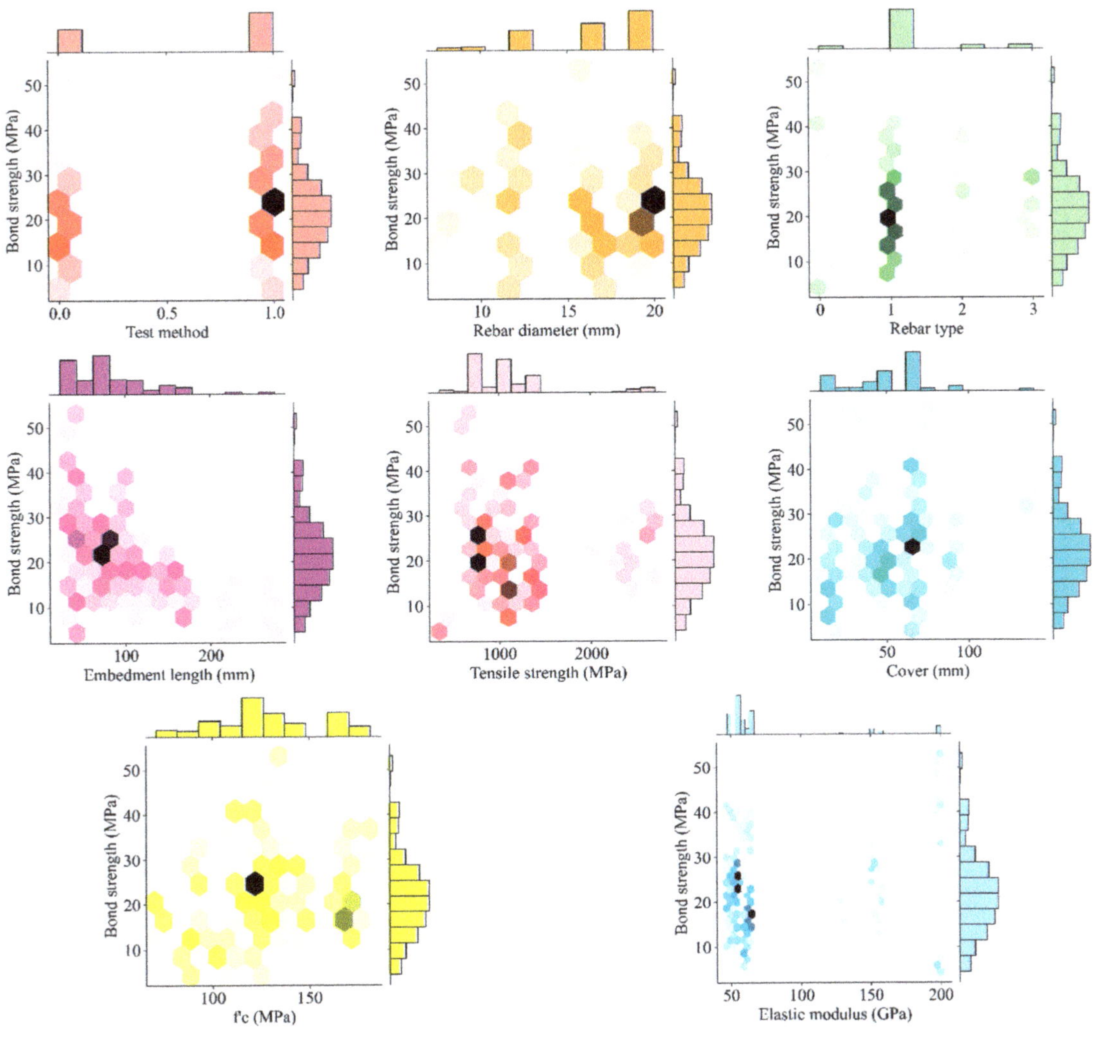

Figure 4. Joint plot output and input variables.

Figure 5 presents a correlation heatmap that visualizes the relationships between input variables and bond strength. The numbers within each heatmap cell represent the correlation coefficient values, which quantify the strength and direction of the relationship between the variables. A correlation coefficient value close to 1 indicates a strong positive correlation, meaning that as one variable increases, the other also increases. A value close to −1 indicates a strong negative correlation, where an increase in one variable corresponds to a decrease in the other. Values near 0 suggest no significant correlation between the variables. These numbers are important for understanding the influence of each input parameter on bond strength, providing insight into which variables are most critical for accurate predictions.

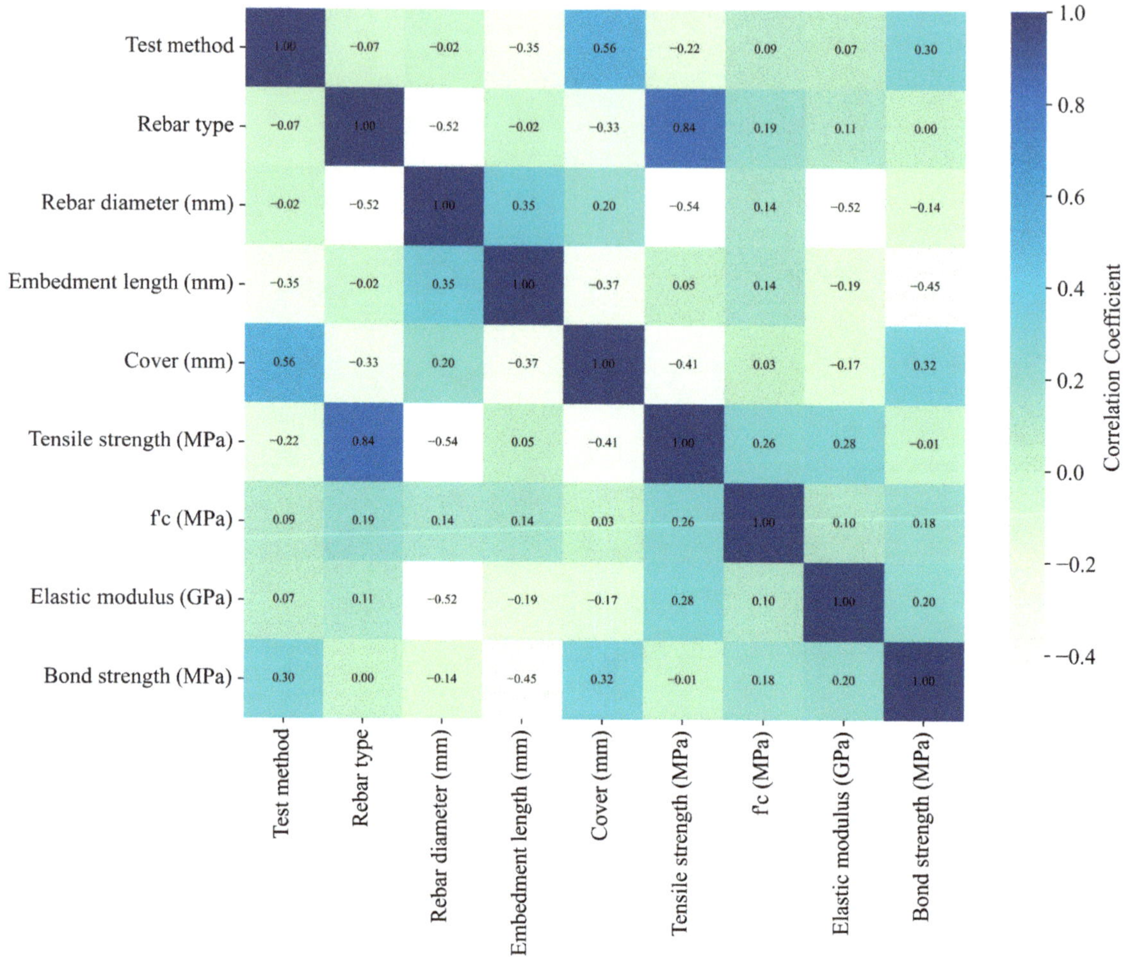

Figure 5. Correlation heatmap between output and input variables.

Figure 6 features histograms and violin plots for both the output and input variables, explicitly focusing on the most impactful features: concrete compressive strength, the tensile strength of rebars, and bond length. The histograms show the distribution of these variables, while the violin plots provide a deeper insight into their distribution characteristics, including density and variability. These visualizations highlight how these key variables are distributed within the dataset and their influence on the bond strength.

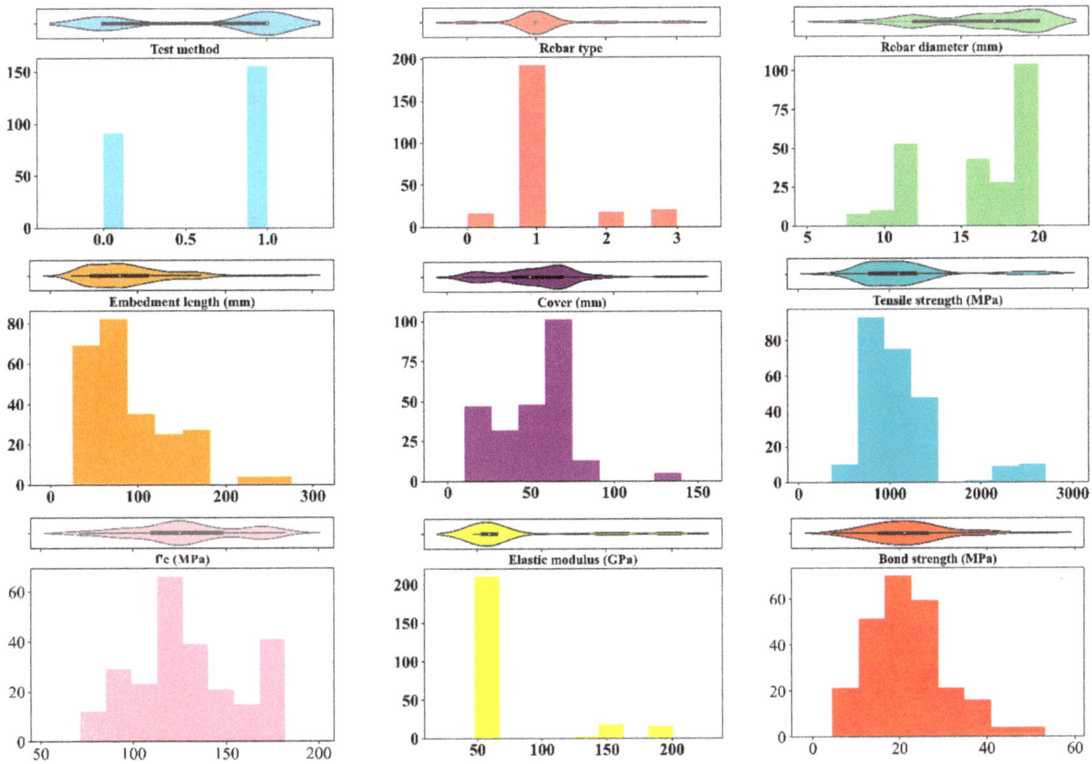

Figure 6. Histograms and violin plot of output and input variables.

5. Boosting

Boosting is an influential ensemble technique in ML that aims to create a strong predictive model by combining the outputs of multiple weak learners, typically decision trees [46]. As shown in Figure 7, the fundamental concept behind boosting is to sequentially train these weak learners so that each new learner focuses on the mistakes made by the previous ones. This iterative process allows the model to gradually improve its accuracy, effectively "boosting" its performance with each step [47]. At the core of boosting lies a simple yet powerful idea: instead of a single, complex model being built, a series of simpler models is constructed, where each successive model is designed to correct the errors of its predecessor. The process begins with a base model often a shallow decision tree, being trained on the entire dataset. The residuals, or errors, from this initial model, are then used to guide the training of the next model in the sequence. Specifically, the subsequent model is trained to predict these residuals, directly addressing the areas where the previous model fell short. This cycle is continued, with each model incrementally refining the predictions made by the ensemble.

AdaBoost is one of the earliest and most popular boosting algorithms. The key idea in AdaBoost is to focus on the instances in which the previous models were misclassified. The Algorithm 1 increases the weights of the misclassified cases so that the subsequent model pays more attention to them [47].

AdaBoost and Gradient Boosting build models sequentially, with each model focusing on correcting the errors of the previous one. However, AdaBoost focuses on misclassification errors, while Gradient Boosting minimizes a specified loss function using gradient descent.

XGBoost and Hist Gradient Boosting are both advanced implementations of Gradient Boosting that focus on improving computational efficiency and accuracy [48]. They incorporate optimizations such as regularization, parallel processing, and efficient data handling, making them faster and more scalable than traditional Gradient Boosting.

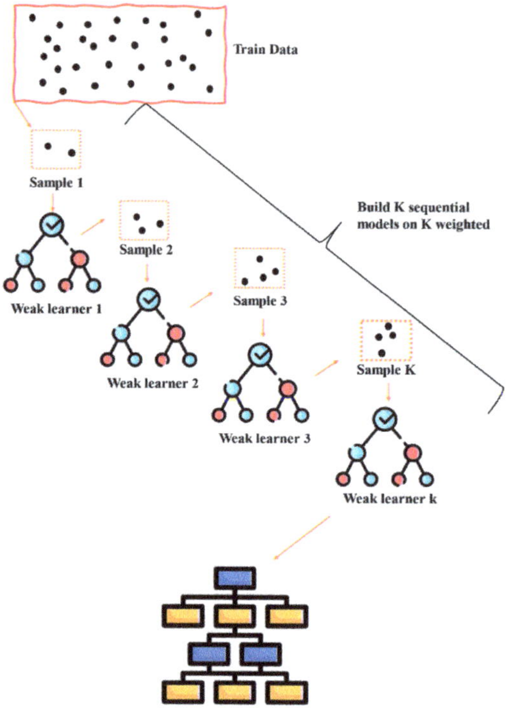

Figure 7. Schematic of boosting algorithm.

Algorithm 1. Algorithm of AdaBoost regressor
STEP 1: Initialize the weight distribution $w_i = \frac{1}{N}\ for\ i = 1, \ldots,\ N$, where N is the number of training samples. STEP 2: For each iteration m: (A) Train a weak learner $h_m(x)$ using the weighted data. (B) Compute the error rate. ϵ_m as: $\epsilon_m = \frac{\sum_{i=1}^{N} w_i . I(y_i \neq h_m(x_i))}{\sum_{i=1}^{N} w_i}$ (C) Compute the model weight α_m: $\alpha_m = log\left(\frac{1-\epsilon_m}{\epsilon_m}\right)$ (D) Update the weights: $w_i \leftarrow w_i.exp(\alpha_m I(y_i \neq h_m(x_i))$ (E) Normalize the weights STEP 3: The final prediction is a weighted majority vote of the weak learners.

CatBoost is specifically designed to handle categorical data more effectively. It introduces ordered boosting, which builds models on subsets of data to prevent overfitting and uses advanced techniques to process categorical features without extensive preprocessing [49]. CatBoost is highly efficient when working with datasets that have a large number of categorical variables. CatBoost is unique among the five models due to its specialized focus on categorical data and unique ordered boosting approach. While it shares the boosting concept with the other models, its techniques and optimizations for categorical features set it apart.

When evaluating the performance of machine learning models, several key metrics are commonly used to assess the accuracy and reliability of predictions. The coefficient of determination, known as R^2, is a statistical measure that represents the proportion of variance in the dependent variable that is predictable from the independent variables, as shown in Equation (2). A higher R^2 value indicates a better fit of the model to the data, with a value of 1 indicating perfect prediction. However, R^2 alone may not always provide a complete picture of model performance, especially in outliers or nonlinear relationships. To complement R^2, as defined in Equation (3), Root Mean Squared Error (RMSE) is often used, providing an absolute measure of the difference between observed and predicted values. RMSE penalizes larger errors more significantly, making it sensitive to outliers. Additionally, Mean Absolute Error (MAE), shown in Equation (4), serves as a robust metric by calculating the average magnitude of prediction errors, regardless of direction, offering a straightforward interpretation of model accuracy. Together, these metrics—R^2, RMSE, and MAE—provide a comprehensive evaluation of model performance, each highlighting different aspects of prediction quality, and are essential for comparing and selecting the most appropriate model for a given task. In these equations, y is the actual value, $\hat{y}$ is the predicted value, and $\overline{y}$ is the mean of the actual values.

$$R^2 = 1 - \frac{\sum_{i=1}^{N} (\hat{y}_{test,i} - y_{test,i})^2}{\sum_{i=1}^{N} (\hat{y}_{test,i} - \overline{y}_{test})^2} \quad (2)$$

$$RMSE = \sqrt{\frac{\sum (\hat{y}_{test,i} - y_{test,i})^2}{N}} \quad (3)$$

$$MAE = \frac{1}{n} \sum_{i=1}^{n} |y_i - \hat{y}_i| \quad (4)$$

6. Hyperparameter Tuning

Hyperparameter tuning is critical in building effective machine-learning models, especially when dealing with complex algorithms such as those used in boosting techniques [50]. The performance of machine learning models heavily depends on the appropriate selection of hyperparameters, which control the behavior of the learning process. Unlike model parameters, which are learned directly from the training data, hyperparameters must be set before the training begins and require careful tuning to optimize model performance.

This study performed hyperparameter tuning using a grid search approach combined with 5-fold cross-validation. The goal was to systematically explore a range of possible hyperparameter values to identify the combination that yields the best performance on the training data while ensuring that the model generalizes well to unseen data.

The flowchart and k-fold cross-validation diagram depict the overall process, as shown in Figures 8 and 9. Initially, the dataset was split into training and testing subsets, with the training set used for model training and hyperparameter tuning and the test set reserved for final model evaluation. For each hyperparameter combination, a model was trained and evaluated using 5-fold cross-validation, where the training data was split into five equally sized folds. Four subsets were used to train the model in each fold, and the remaining subset was used for validation. This process was repeated five times, with each subset serving as the validation set once, and the average performance metric (R^2 score) across the folds was computed.

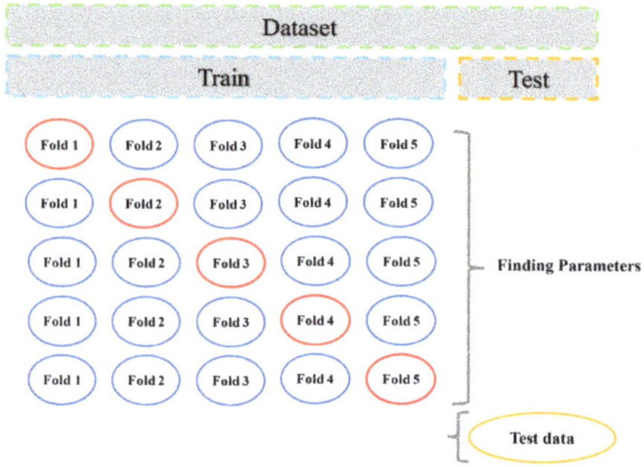

Figure 8. Schematic of 5-fold cross validation (Blue folds are training data and red folds are validation data).

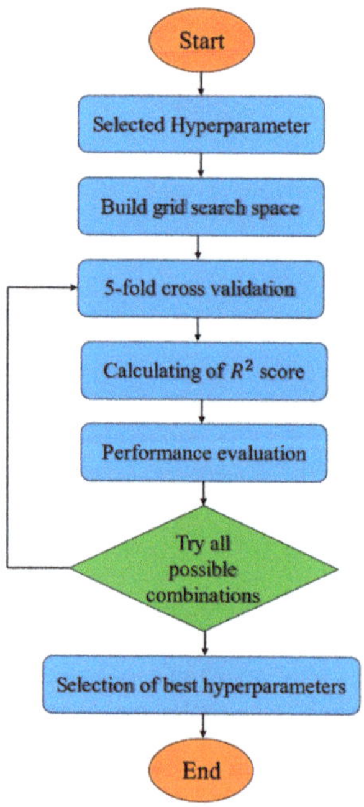

Figure 9. Flowchart of grid search.

Upon completion of the grid search, the best combination of hyperparameters was identified based on the highest mean R^2 score obtained during cross-validation. This optimal set of hyperparameters was then used to train the final model on the entire training set. The model's performance was evaluated before and after tuning to assess the impact of hyperparameter optimization. The selected hyperparameters and their space for hyperparameter tuning are shown in Tables 3–6.

Table 3. Adaboost selected hyperparameters and their space.

N_Estimators	Learning_Rate	Loss
50	0.01	Linear
100	0.1	Square
200	0.2	Exponential
300	0.3	-
400	0.5	-

Table 4. Catboost selected hyperparameters and their space.

Iterations	Learning_Rate	Depth	L2_Leaf_Reg	Bagging_Temperature
100	0.01	4	3	0.8
150	0.05	6	5	1
200	0.1	8	7	-
300	0.2	-	9	-
400	-	-	-	-

Table 5. Gradient booting and XGBoost selected hyperparameters and their space.

N_Estimators	Learning_Rate	Max_Depth	Max_Features
50	0.01	None	sqrt
100	0.1	4	log2
200	0.2	5	-
300	0.3	8	-
400	0.5	10	-

Table 6. Hist Gradient Boosting Regressor selected hyperparameters and their space.

L2_Regularization	Learning_Rate	Max_Depth	Max_Iter
0	0.01	None	100
0.1	0.1	4	200
0.5	0.2	5	300
1	0.3	8	-
-	0.5	10	-

7. Results

7.1. Machine Learning Results

The performance evaluation of the ML models was conducted using three key metrics: R^2 score, RMSE, and MAE. These metrics were calculated both before and after hyperparameter tuning to assess the impact of the tuning process on model accuracy.

As presented in Table 7 and Figure 10, the results reveal that hyperparameter tuning significantly enhanced the performance of all models, particularly those that initially exhibited lower accuracy.

Table 7. ML models' results.

Model	R^2 (Default)	RMSE (Default)	MAE (Default)	R^2 (Tuned)	RMSE (Tuned)	MAE (Tuned)
ADAboost	0.61	6.46	5.03	0.7	5.63	2.67
Catboost	0.94	2.56	1.9	0.95	2.34	1.74
Gradient Boosting	0.94	2.58	1.95	0.95	2.26	1.73
XGBoost	0.94	2.33	1.78	0.95	2.21	1.68
Hist Gradient Boosting	0.59	6.58	4.18	0.68	5.86	3.24

After tuning, the AdaBoost model's R^2 score improved from 0.61 to 0.7, indicating a better fit between the predicted and actual values. This improvement was accompanied by a reduction in RMSE from 6.46 to 5.63 and a decrease in MAE from 5.03 to 2.67, demonstrating that tuning effectively reduced the model's prediction errors.

CatBoost, which already performed well with default parameters, saw its R^2 score increase slightly from 0.94 to 0.95 after tuning. The RMSE decreased from 2.56 to 2.34, and the MAE was reduced from 1.9 to 1.74. Although the improvements were marginal, they indicate that even highly effective models can benefit from careful tuning.

Similarly, after tuning, the Gradient Boosting model slightly increased its R^2 score from 0.94 to 0.95. The RMSE improved from 2.58 to 2.26, and the MAE decreased from 1.95 to 1.73. These results suggest that while the model was already robust, hyperparameter tuning contributed to further refining its predictions.

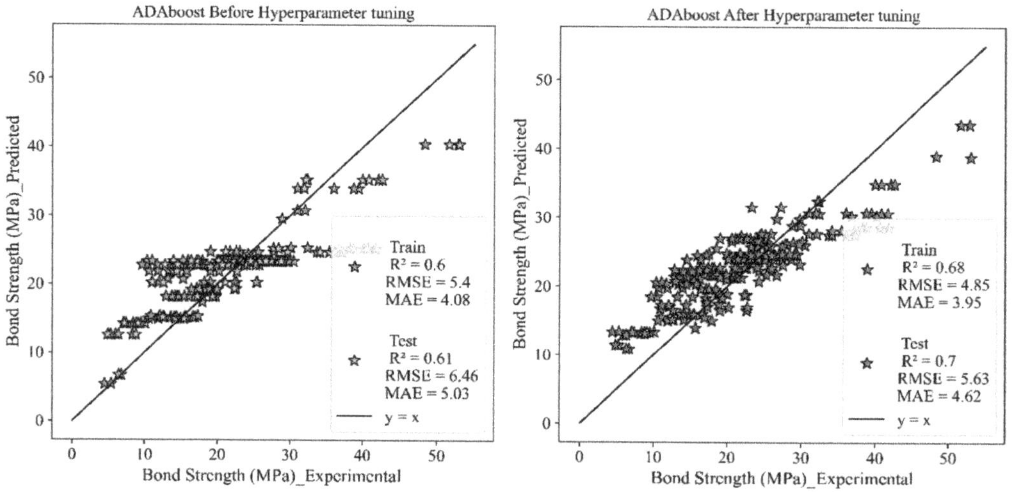

(a)

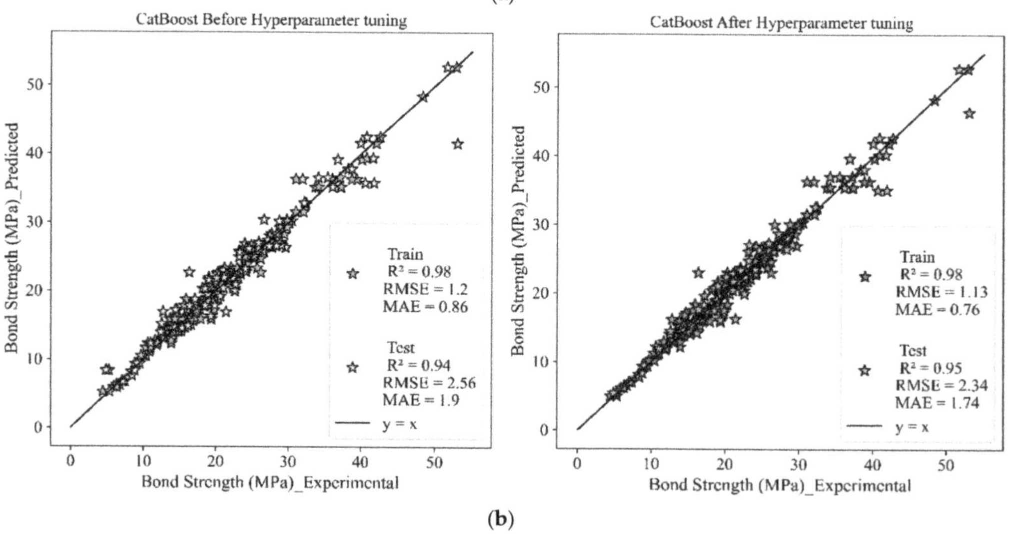

(b)

Figure 10. *Cont.*

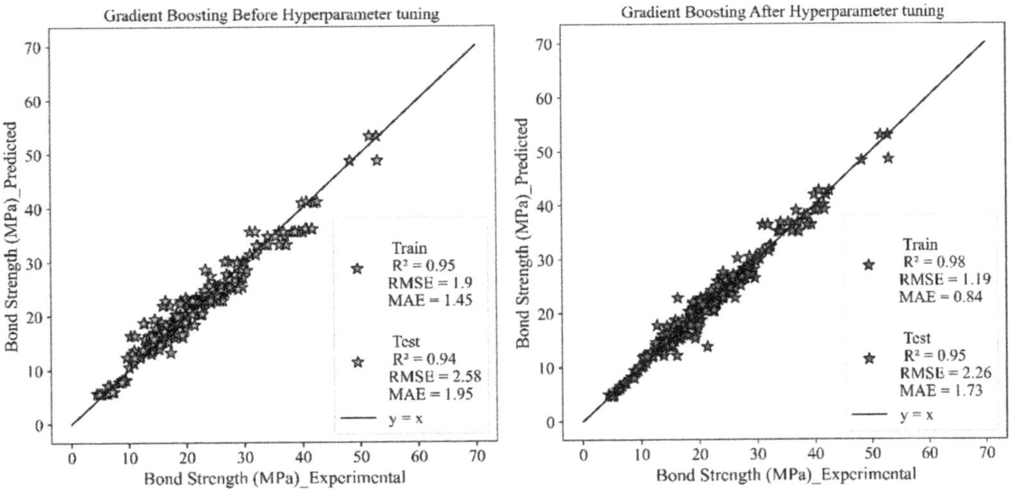

(**c**)

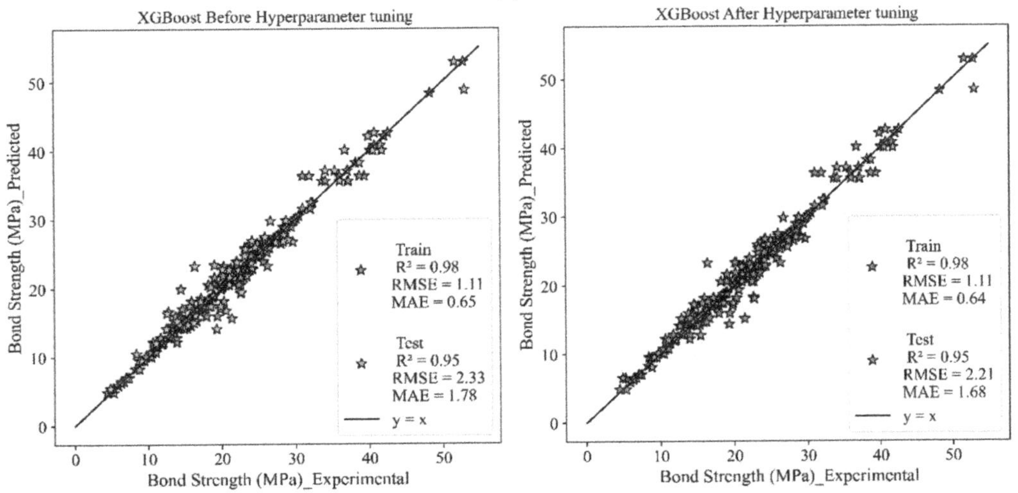

(**d**)

Figure 10. *Cont.*

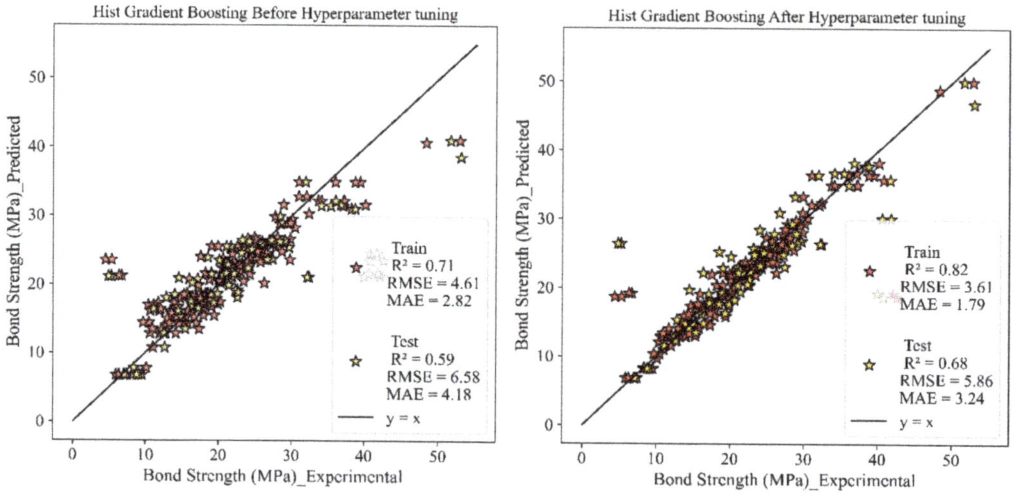

(e)

Figure 10. ML model results: (**a**) ADAboost, (**b**) CatBoost, (**c**) Gradient Boosting, (**d**) XGBoost, and (**e**) Hist Gradient Boosting.

XGBoost, known for its high performance, also showed a modest improvement in accuracy after tuning, with the R^2 score increasing from 0.94 to 0.95. The RMSE dropped from 2.33 to 2.21 and the MAE from 1.78 to 1.68, indicating a slight enhancement in the model's predictive capabilities.

The Hist Gradient Boosting model observed the most notable improvement, where the R^2 score significantly increased from 0.59 to 0.68 after tuning. Although the final R^2 score of 0.68 does not reach the same level as the other models, the improvement is still substantial, indicating that hyperparameter tuning was crucial in enhancing the performance of this model. The tuning process also reduced RMSE from 6.58 to 5.86 and the MAE from 4.18 to 3.24, further demonstrating the positive impact of optimization on the model's predictive accuracy.

The impact of hyperparameter tuning is shown in Figure 11, which clearly demonstrates that hyperparameter tuning plays a crucial role in enhancing model performance. The improvements were especially pronounced for models such as AdaBoost and Hist Gradient Boosting, which initially had lower R^2 scores. After tuning, these models achieved higher R^2 scores and exhibited lower RMSE and MAE values, significantly reducing prediction errors. Even models that performed well with default parameters, such as CatBoost and XGBoost, benefited from tuning, achieving slight but meaningful improvements in accuracy. Also, the best hyperparameter values resulting from hyperparameter tuning are shown in Table 8.

Table 8. Best hyperparameters resulting from hyperparameter tuning.

Model	N_Estimators	Max_Depth	Max_Features	Learning_Rate	Loss
AdaBoost	100	-	-	0.5	exponential
Gradient Boosting	200	4	sqrt	0.1	-
XGBoost	150	10	-	0.1	-

These findings underscore the necessity of hyperparameter optimization in developing reliable and accurate machine learning models, particularly in complex applications such as structural engineering. The consistent performance gains across all models suggest that thorough hyperparameter tuning is essential for fully leveraging the potential of machine learning algorithms and achieving optimal results.

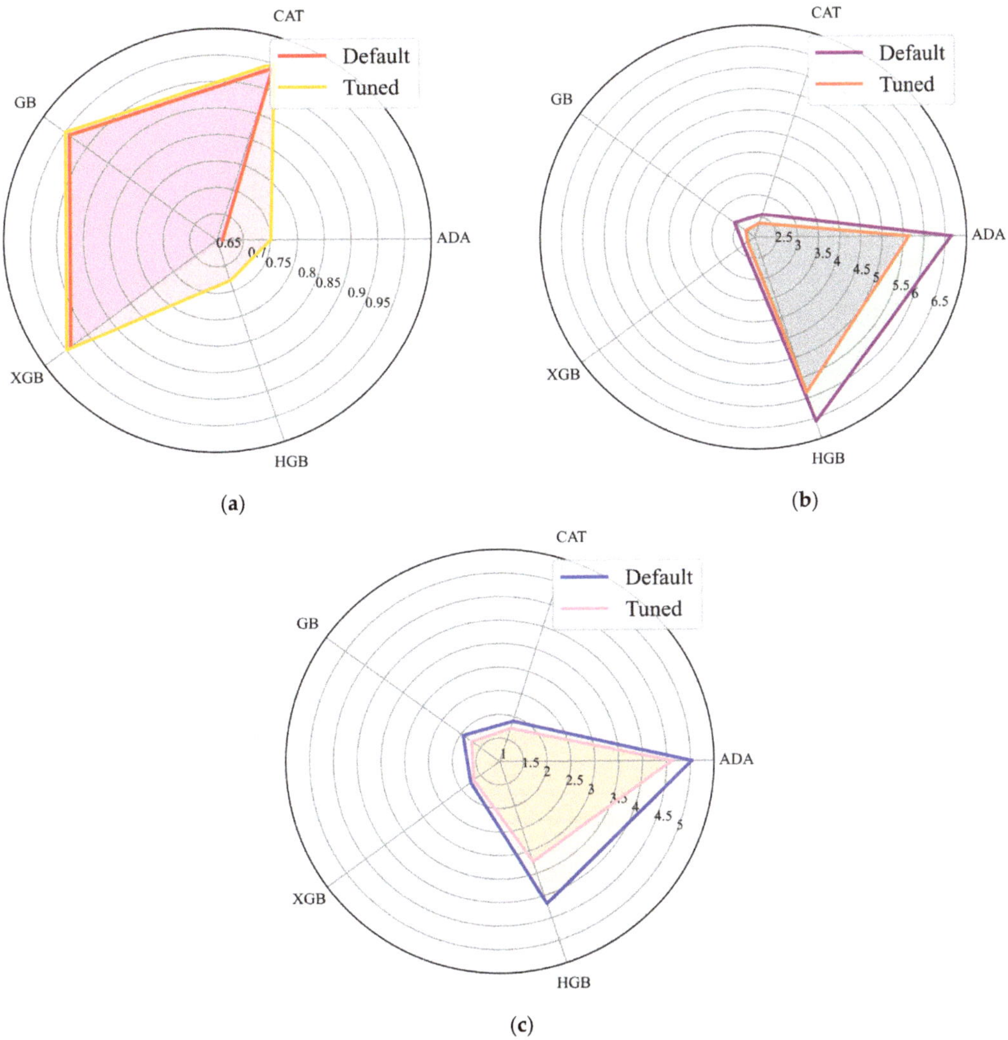

Figure 11. Impact of hyperparameter tuning on ML models results: (**a**) R^2, (**b**) RMSE, and (**c**) MAE.

The Taylor diagrams in Figure 12 visually compare the ML models' performance on the training, test, and combined datasets. Across all datasets, CatBoost, Gradient Boosting, and XGBoost consistently demonstrate high correlation coefficients and standard deviations that closely match the reference, indicating their robustness and accuracy in capturing the underlying patterns of the data. AdaBoost shows moderate performance, with slightly lower correlations and greater deviations from the reference, while Hist Gradient Boosting, despite improvements after hyperparameter tuning, still exhibits lower correlation and a higher standard deviation compared to the other models. These diagrams highlight the effectiveness of CatBoost, Gradient Boosting, and XGBoost in delivering reliable predictions while also pointing to areas where models such as AdaBoost and Hist Gradient Boosting could benefit from further refinement.

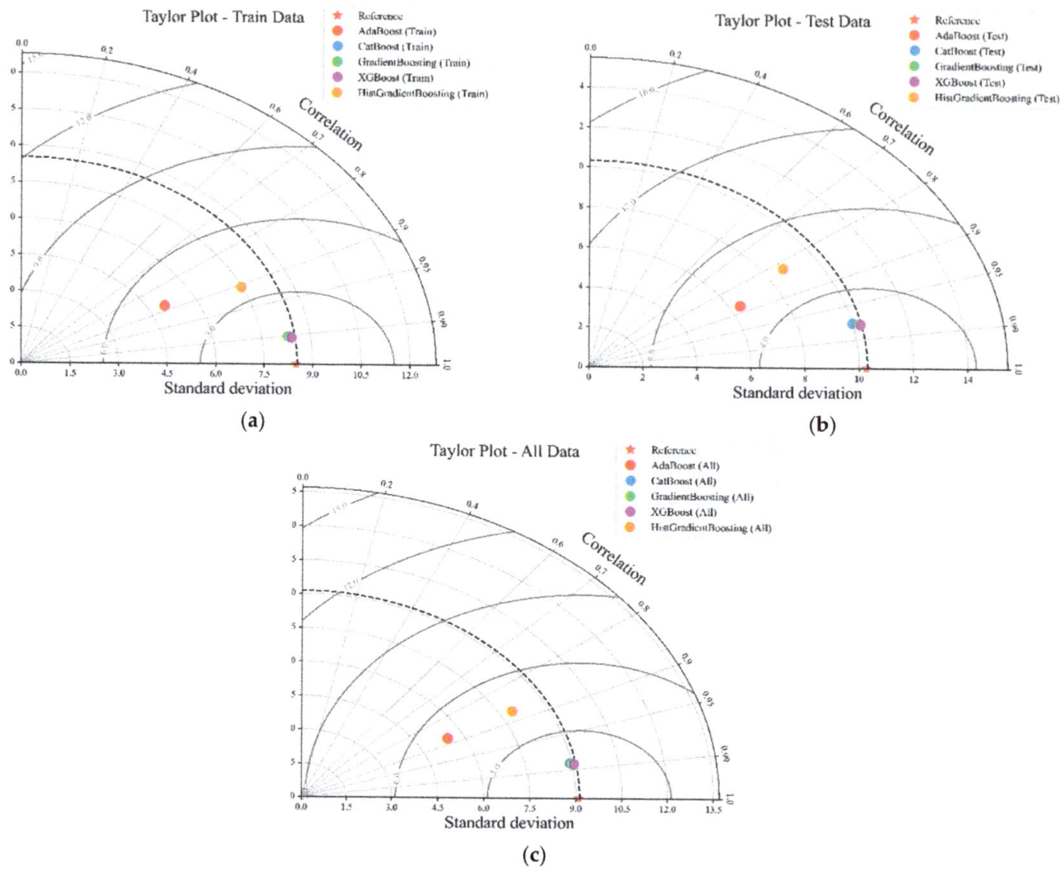

Figure 12. Taylor diagram for ML models results: (**a**) train data, (**b**) test data, and (**c**) total data.

To further enhance the model performance, two approaches were explored using a Voting Regressor, which combines predictions from multiple models to leverage the strengths of each.

In the first approach, the Voting Regressor was constructed by combining all the models: AdaBoost, CatBoost, Gradient Boosting, XGBoost, and Hist Gradient Boosting. The result of this ensemble was comparable to the performance of the best individual models (CatBoost, Gradient Boosting, and XGBoost), with no significant improvement in the key metrics (see Figure 13a). This suggests that while combining all models can help in averaging out errors, it does not necessarily lead to better performance if some of the models are less accurate.

The Voting Regressor was formed in the second approach using only the best-performing models: CatBoost, Gradient Boosting, and XGBoost. This targeted ensemble approach resulted in a slight performance improvement, with the R^2 score increasing from 0.95 (achieved by the best individual models) to 0.96 (see Figure 13b). This modest improvement indicates that focusing the ensemble on the top-performing models allows the Voting Regressor to deliver more accurate and consistent predictions by capitalizing on the strengths of these models without the dilution effect that might come from including weaker models.

Overall, the selective combination of the best models in the Voting Regressor proved to be a more effective strategy, providing a small but valuable boost in predictive accuracy further enhancing the model's reliability in predicting bond strength in structural engineering applications.

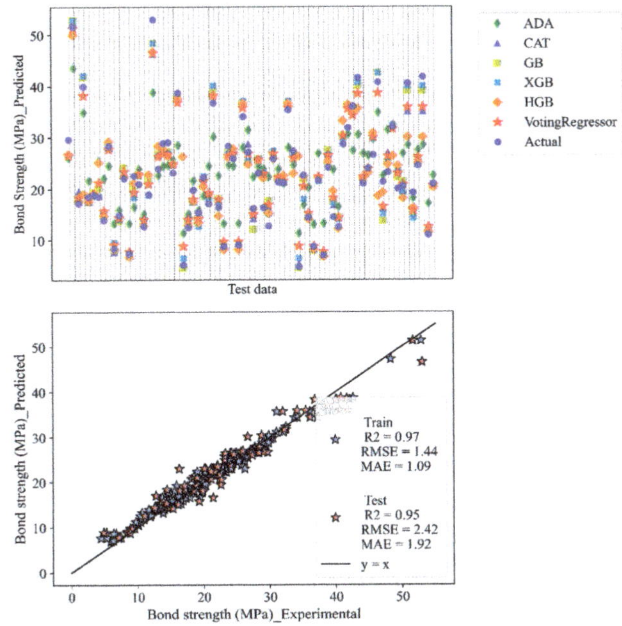

(a)

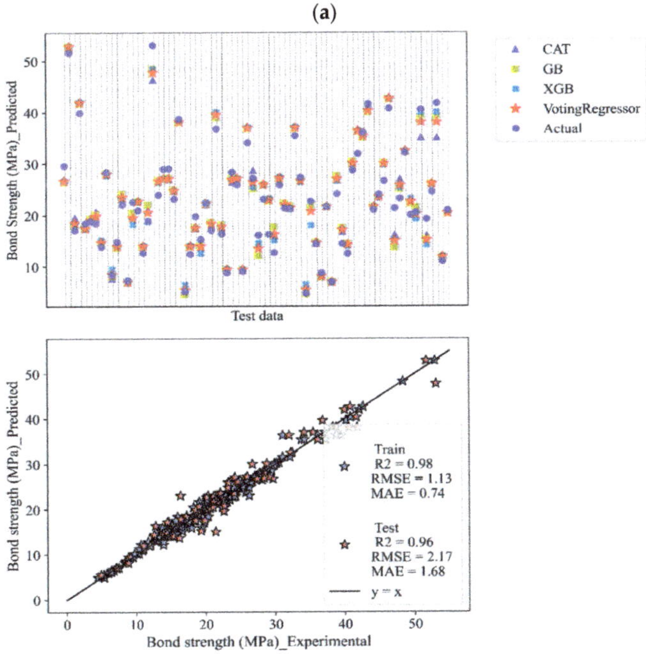

(b)

Figure 13. Voting regressor results: (**a**) first approach and (**b**) second approach.

Additionally, the performance of all models across all metrics, including R^2, RMSE, and MAE, is shown in Figure 14.

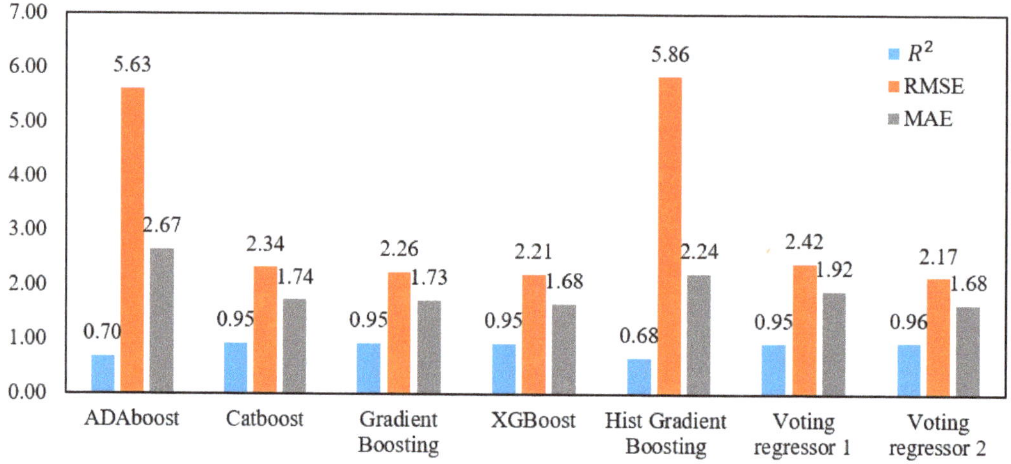

Figure 14. Comparison of the results of all ML models.

The developed user interface (UI) in this study provides a comprehensive and interactive platform for implementing machine learning models tailored explicitly for bond strength prediction in FRP-reinforced UHPC. As shown in Figure 15, this UI allows users to select from six different machine learning models, which are discussed, each with its own specific set of hyperparameters. The interface enables users to input various features related to the structural properties of the FRP-UHPC system and customize the model parameters. After model training, users can evaluate the model's performance using key metrics such as R^2, RMSE, and MAE, displayed for both the training and test datasets. Furthermore, the UI facilitates the prediction of bond strength by allowing users to input new data based on the trained model, thus offering an accessible and powerful tool for practical applications and research in structural engineering. The Python code for this UI can be found on GitHub.

The feature importance analysis across various ML models provides critical insights into the factors that most significantly influence the prediction of bond strength. As shown in Figure 16, each model highlights different aspects of the input features, allowing for a deeper understanding of the variables that drive the predictions.

In the AdaBoost model, tensile strength (MPa), elastic modulus (GPa), and embedment length (mm) are identified as the most influential features. These variables dominate the model's decision-making process, emphasizing their critical role in predicting bond strength. Other features such as cover and compressive strength (fc) also contribute, but their impact is less pronounced.

The CatBoost model recognizes tensile strength (MPa) and embedment length (mm) as key predictors. Features such as cover (mm) and elastic modulus (GPa) also show significant importance, reflecting the model's sensitivity to these parameters. CatBoost's advanced handling of categorical variables may account for the subtle differences in feature importance distribution compared to other models.

Gradient Boosting highlights embedment length (mm) as the most critical feature, followed closely by tensile strength (MPa) and elastic modulus (GPa). This emphasis on embedment length aligns with established engineering principles, reinforcing its importance in determining bond strength.

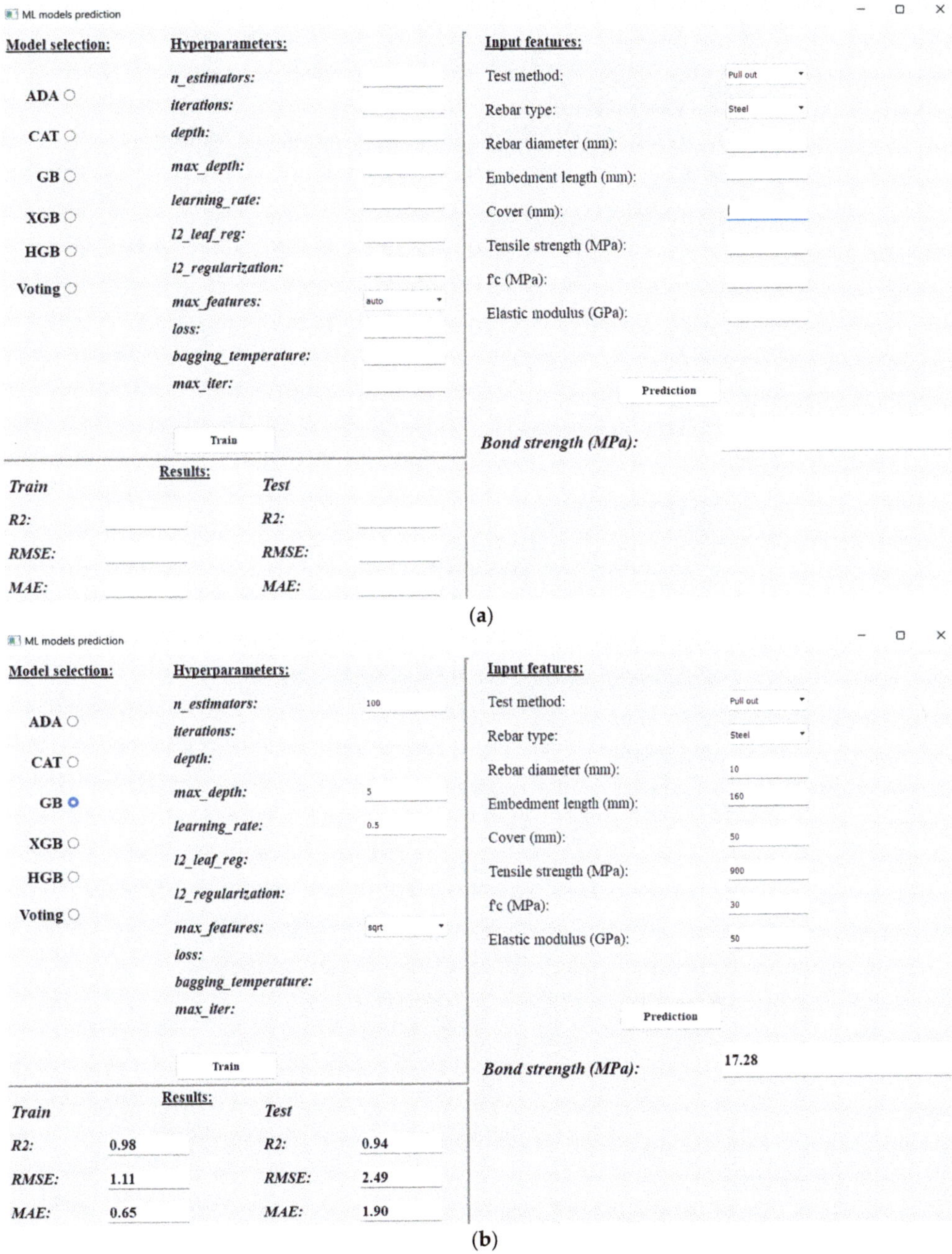

Figure 15. ML models UI: (**a**) before running and (**b**) after running.

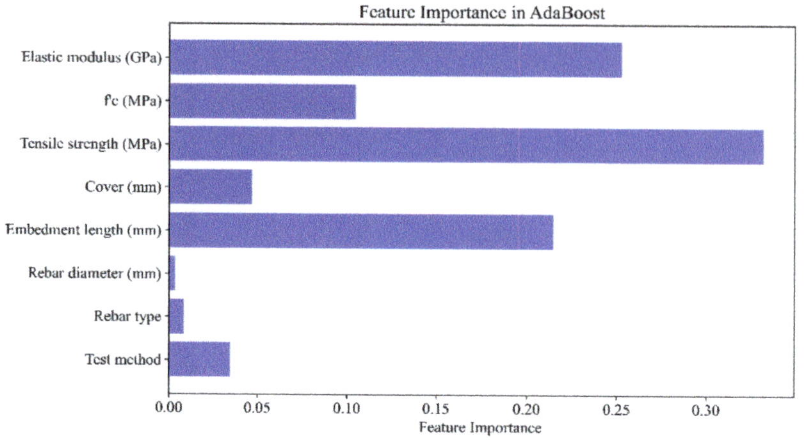

(**a**)

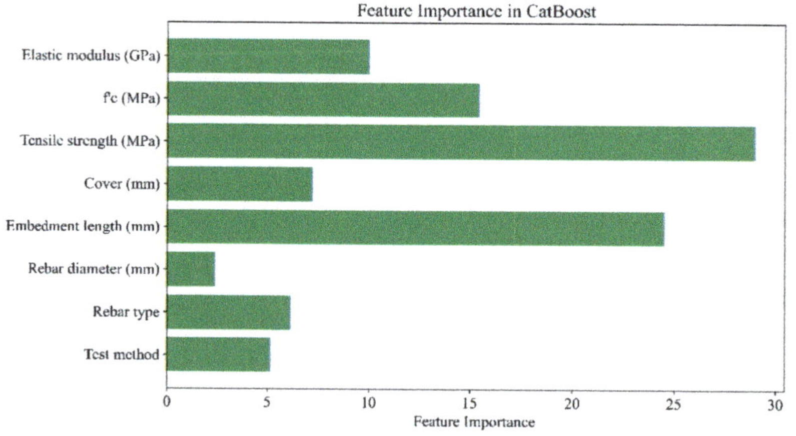

(**b**)

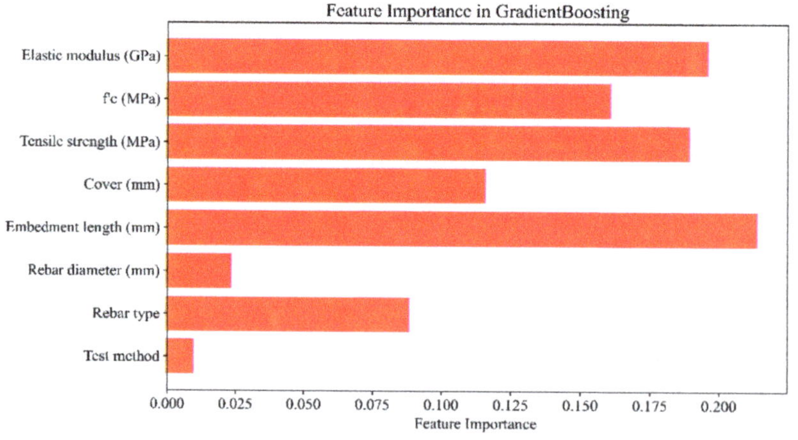

(**c**)

Figure 16. *Cont.*

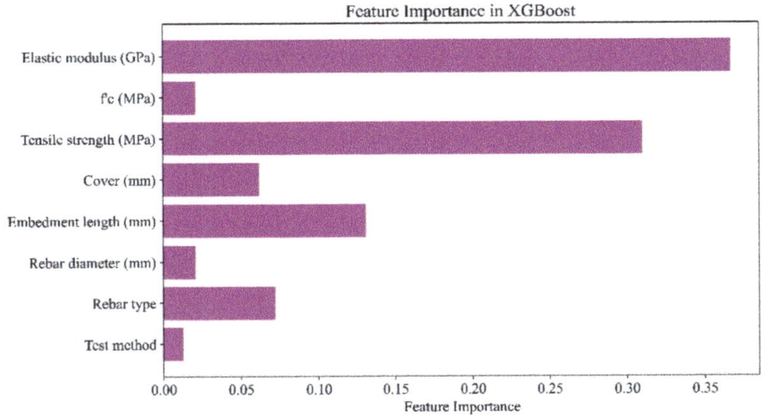

(**d**)

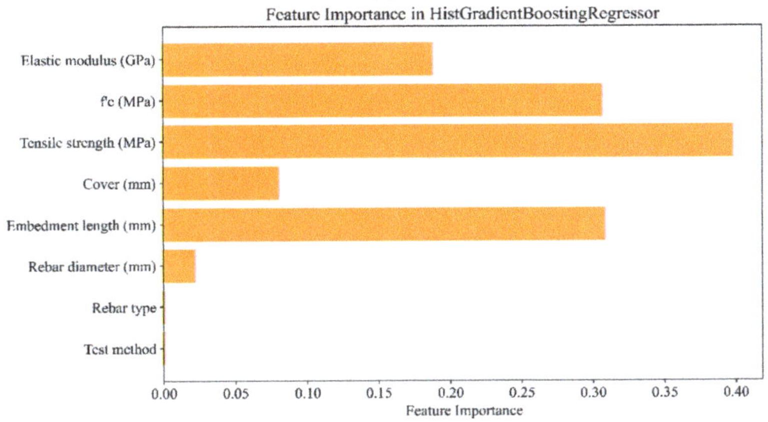

(**e**)

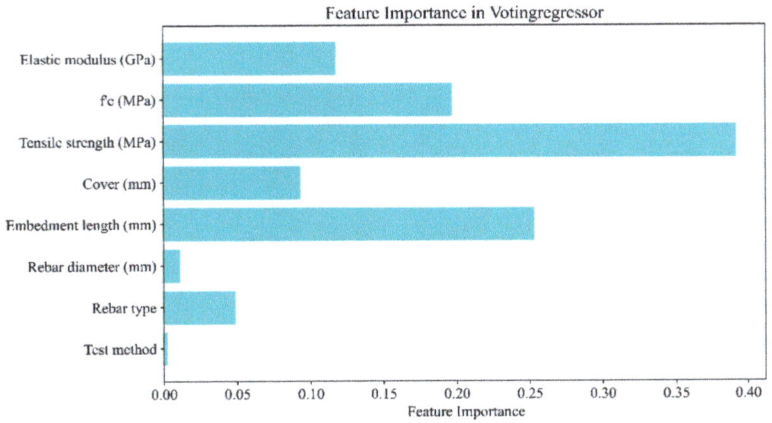

(**f**)

Figure 16. Feature importance of ML models: (**a**) Adaboost, (**b**) CatBoost, (**c**) Gradient Boosting, (**d**) XGBoost, (**e**) Hist Gradient Boosting, and (**f**) Voting Regressor.

XGBoost places the highest importance on elastic modulus (GPa) and tensile strength (MPa), with embedment length (mm) also playing a significant role. The distribution of feature importance in XGBoost reflects its unique optimization techniques, influencing how the model prioritizes variables.

In Hist Gradient Boosting, embedment length (mm) and tensile strength (MPa) emerge as the top features, with elastic modulus (GPa) also being crucial. The model's use of histogram-based binning may contribute to how it evaluates and prioritizes features, leading to a slightly different emphasis than other boosting models.

The Voting Regressor, which combines the predictions from multiple models, consistently identifies tensile strength (MPa) and embedment length (mm) as the most important features. This consistency across different models underscores the critical influence of these parameters in predicting bond strength. By balancing the feature importance from all its constituent models, the Voting Regressor offers a more comprehensive understanding of the factors driving bond strength predictions.

As a result, tensile strength (MPa), elastic modulus (GPa), and embedment length (mm) are consistently recognized as the key predictors of bond strength across all models. This consistency highlights the robustness of these features in both modeling and practical engineering applications. The variations in feature importance among the models also demonstrate the value of using an ensemble approach such as the Voting Regressor, which captures a more nuanced and balanced understanding of feature contributions, ultimately leading to more accurate and reliable predictions.

7.2. Shapley Values

In addition to traditional feature importance analysis, SHAP values were employed to gain a more nuanced understanding of the impact of each feature on the model's predictions. As shown in Figure 17, SHAP values offer a method to explain the output of a machine-learning model by attributing the contribution of each feature to the final prediction. This approach is rooted in cooperative game theory, where the goal is to fairly distribute the "payout" (in this case, the model's prediction) among all features based on their individual contributions.

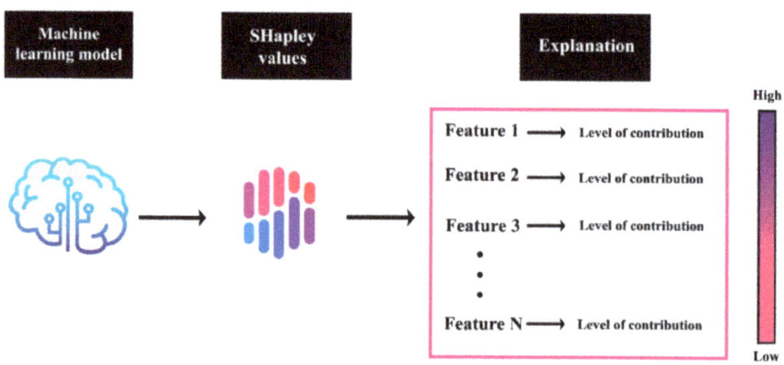

Figure 17. Workflow of the Shapley values method.

$$\phi_i = \sum_{S \subsetneq N\{i\}} \frac{|s|!(|N| - |S| - 1)!}{|N|!} [v(S \cup \{i\}) - v(S)] \quad (5)$$

As shown in Equation (5), the Shapley value ϕ_i for feature i is determined by averaging the marginal contributions of that feature across all possible permutations of features. Here, N represents the set of all features, S denotes a subset of features excluding feature i, $|S|$ indicates the number of elements in subset S, $v(S)$ is the model's prediction based solely on the features in S, and $v(S \cup \{i\})$ is the model's prediction when feature i is included in S.

Figures 18 and 19 illustrate the SHAP value analysis for the XGBoost model, which was identified as one of the best-performing models in this study. In Figure 18, the SHAP summary plot shows the distribution of SHAP values for each feature across all predictions. Each point on the plot represents a SHAP value for a particular feature and instance. The color represents the feature value, and the position on the x-axis shows the SHAP value, indicating whether the feature increases or decreases the predicted bond strength. This plot provides a detailed view of how each feature affects individual predictions, highlighting the variability in the impact of features such as embedment length and tensile strength.

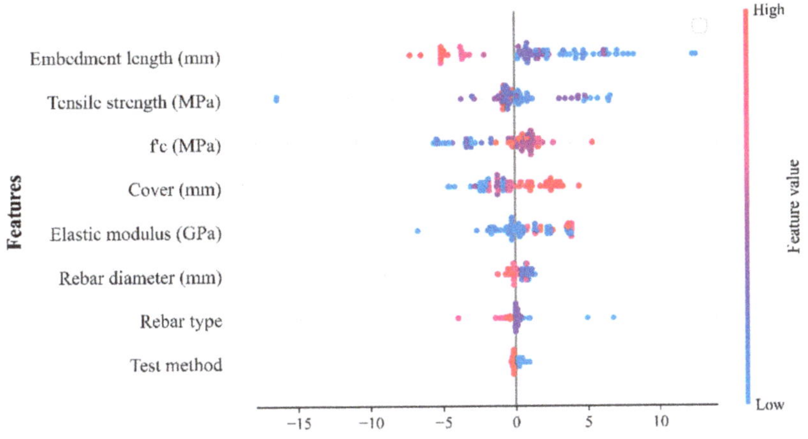

Figure 18. SHAP values for the XGB model.

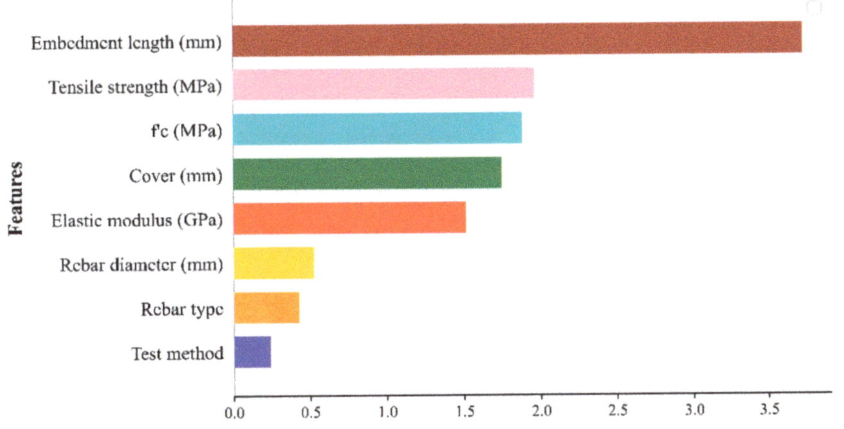

Figure 19. Mean SHAP values for the XGB model.

Figure 19 presents the importance of global SHAP features, where features are ranked by their average absolute SHAP values. This gives an overview of which features have the most significant overall impact on the model's predictions. Interestingly, embedment length (mm) is ranked as the most important feature in the SHAP analysis, while in the traditional feature importance analysis, it is ranked third. This discrepancy arises because traditional feature importance measures the average contribution of each feature to the model's overall accuracy, whereas SHAP values take into account the impact of each feature on every single prediction.

The higher ranking of embedment length in the SHAP analysis suggests that while its average contribution might be lower compared to features such as elastic modulus and tensile strength, its impact is more significant and variable in certain contexts. This variability could mean that, in specific cases, changes in embedment length have a more substantial effect on the model's output, thereby increasing its overall importance when assessed through SHAP values. Essentially, SHAP values capture the feature's influence more granularly, reflecting its critical role in specific instances rather than just its average contribution.

The primary difference between SHAP values and traditional feature importance lies in the interpretation and granularity of the analysis. Traditional feature importance measures how much each feature contributes to the model's predictions on average, but it does not account for the direction or variability of these contributions. In contrast, SHAP values provide a more detailed explanation by showing the magnitude of a feature's impact and the direction (positive or negative) and how this impact varies across different instances. This makes SHAP values particularly useful for understanding complex models such as XGBoost, where interactions between features can lead to varying impacts on predictions.

To summarize, embedment length ranks as the most important feature in SHAP analysis on the test data because it has a significant and context-specific impact on individual predictions, even if its overall contribution across the entire training dataset (as measured by traditional feature importance) is somewhat lower. This highlights the value of using SHAP values to understand feature importance better, especially when analyzing how the model performs on new, unseen data.

Figure 20 provides a detailed visualization of the SHAP values for individual features in the XGBoost model, highlighting how each feature impacts the model's predictions on the test data. Each subplot corresponds to a specific feature, with the SHAP values plotted on the y-axis and the feature values on the x-axis, while the color gradient, from blue to red, represents the range of feature values. For categorical features such as Test Method and rebar type, distinct clusters of SHAP values are observed, indicating how specific categories within these features consistently influence the predicted bond strength, either positively or negatively. Rebar diameter (mm) shows that higher diameters generally result in higher SHAP values, suggesting a positive correlation with bond strength. However, the effect varies depending on the interaction with other features. Embedment length (mm) significantly impacts predictions, where shorter lengths tend to decrease predicted bond strength and longer lengths have a positive effect, illustrating this feature's critical role in the model's output. The feature Cover (mm) demonstrates a more complex, nonlinear relationship, where increases in cover can either positively or negatively affect the predictions, reflecting the nuanced role of this feature. Lastly, tensile strength shows a clear trend where higher tensile strength leads to higher predicted bond strength, further reinforcing its importance in the model. Figure 21 presents SHAP waterfall plots for three specific instances from the test dataset (numbers 14, 25, and 69) to illustrate how individual features contribute to the final model prediction in the XGBoost model. Each plot shows the breakdown of the model's prediction and the contributions from each feature, offering a clear visualization of how the features interact to influence the predicted bond strength.

7.3. Predictive Formulas for Bond Strength of FRP Rebars in UHPC

There have been numerous attempts to predict the bond strength of FRP rebars in UHPC, with each study incorporating specific FRP rebars and concrete characteristics. Table 9 summarizes some of these predictive formulas derived from various research efforts.

Figure 22 presents the results of applying these predictive formulas to the compiled dataset, with comparisons to the corresponding experimental values. These formulas are overfitted to the specific conditions of their original research papers and do not generalize well to the broader dataset. This overfitting limits their applicability when predicting bond strength in more diverse scenarios.

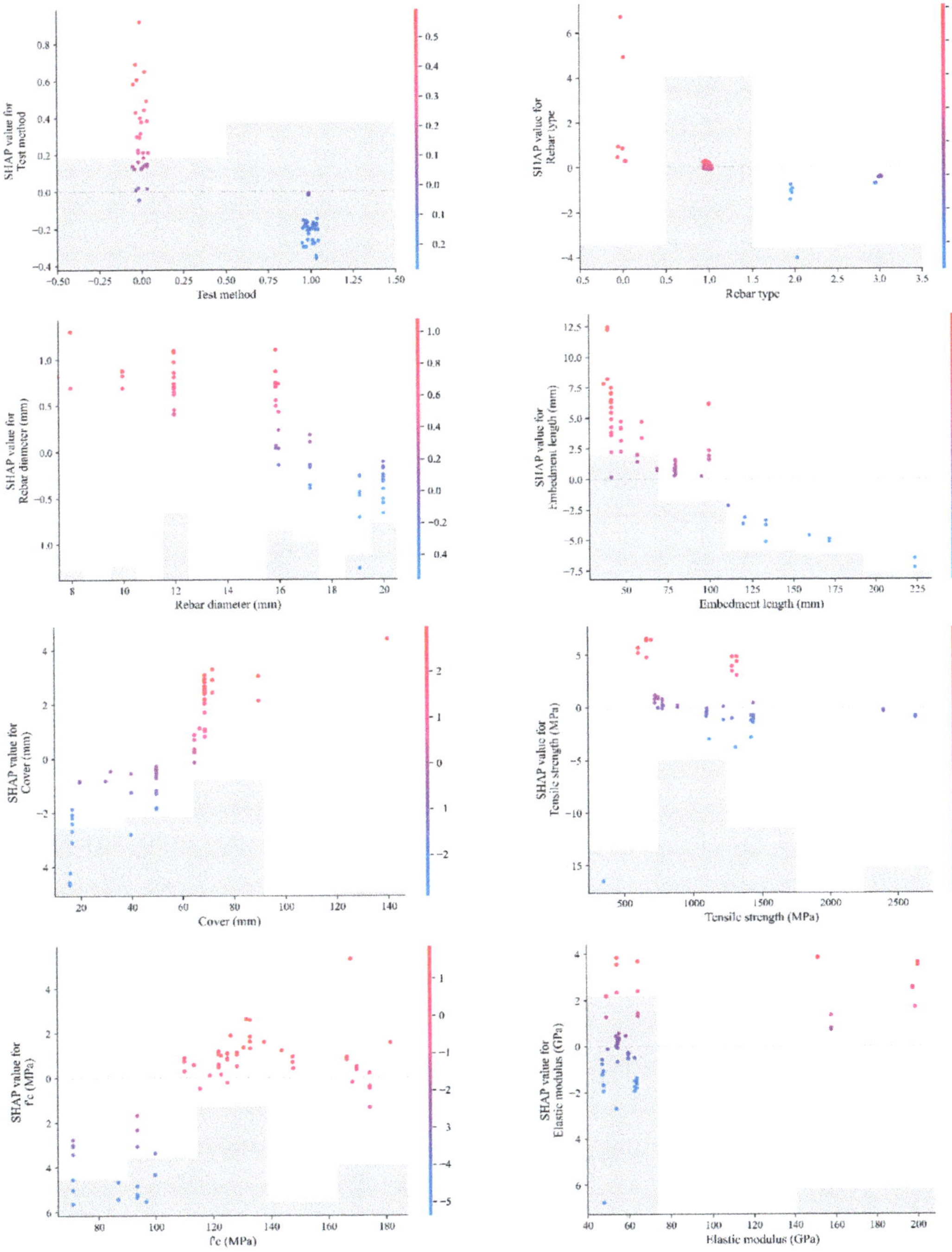

Figure 20. SHAP values for each feature.

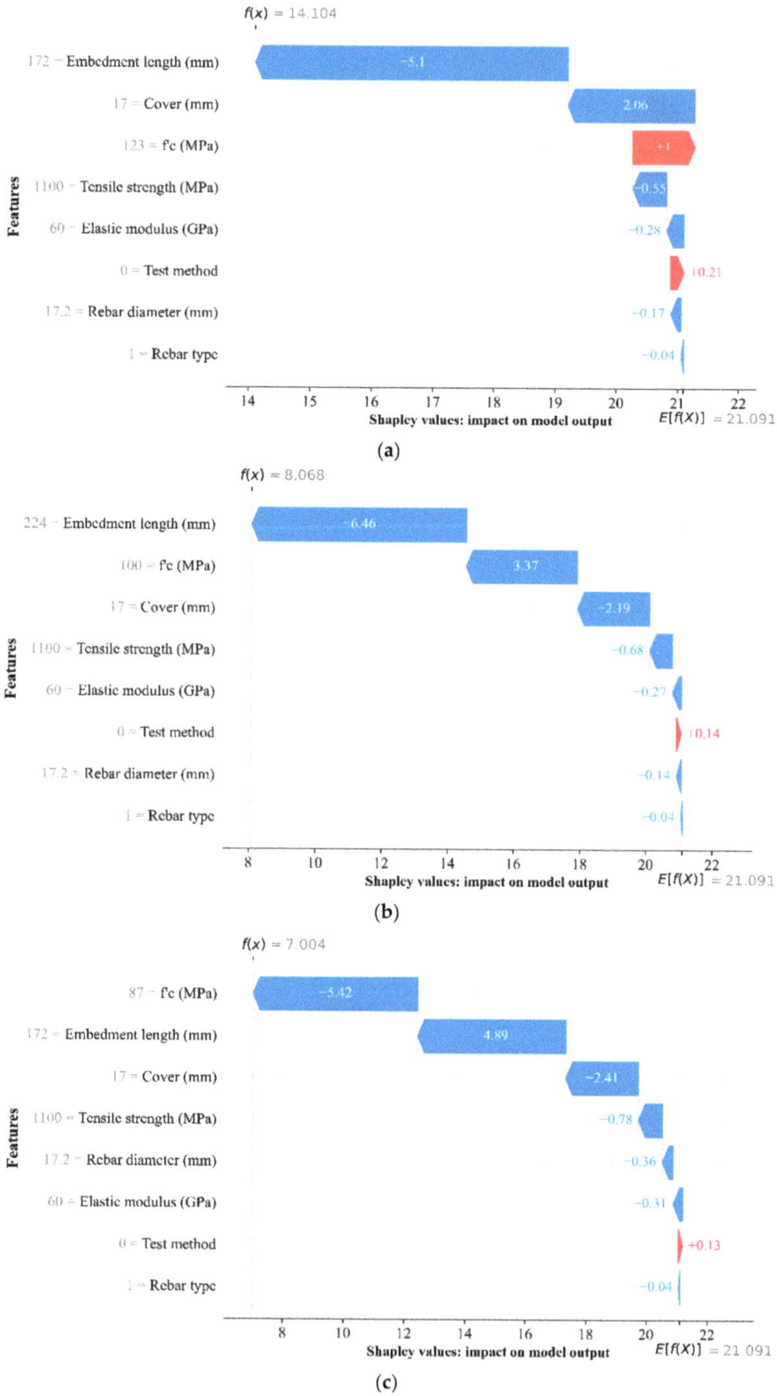

Figure 21. SHAP values for three random points in test data: (**a**) test data number 19, (**b**) test data number 27, and (**c**) test data number 54 (In the visualization, blue color indicates that the feature reduces the prediction value, while red color shows that the feature increases the prediction value).

Table 9. Proposed formulations in previous studies.

Research	Formula	Note
Yoo et al. 2023 [8]	(a): $\frac{\tau_{max}}{(f_c')^{0.36}} = (1.46 + 0.043d_b) \times (0.68 + 0.195\frac{c}{d_b} + 2.449\frac{d_b}{l_e})$ for sand-coated CFRP (b): $\frac{\tau_{max}}{(f_c')^{0.5}} = (7.775 + 1.184d_b) \times \left(-0.918 - 0.061\frac{d_b}{l_e} + \left(\frac{c}{d_b}\right)^{0.03}\right)$ for helically ribbed CFRP	Separate equation according to CFRP type.
Lee et al. 2008 [51]	$\tau_{max} = 3.3f_c'^{0.3}$ for GFRP bars (a) $\tau_{max} = 4.1f_c'^{0.5}$ for steel bars (b)	Only considers f_c'.
Zhu et al. [24]	$\tau_{max} = \left(-0.5 + 0.03\frac{c}{d_b} + 4.5\frac{d_b}{l_e} + \frac{21.6}{d_b}\right)\sqrt{f_c'}$ CFRP rebars in UHPC based on both pullout test data and beam test	Does not consider rebar type.

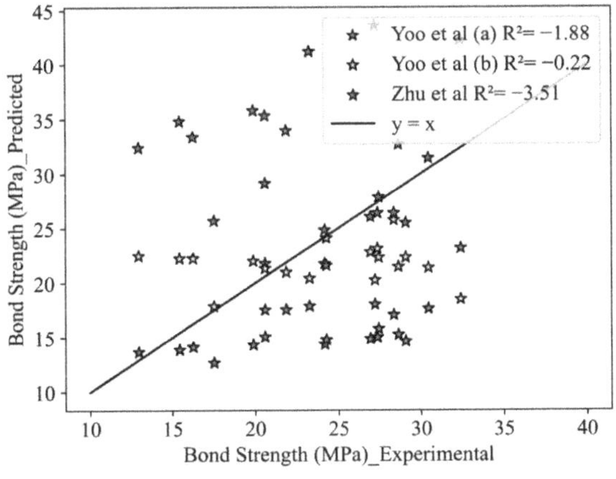

(**a**)

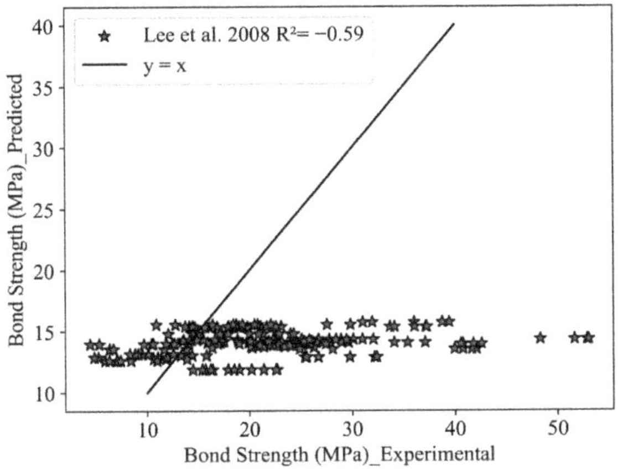

(**b**)

Figure 22. Previous predictive formulas compare to experimental values: (**a**) CFRP and (**b**) GFRP and steel [8,24,51].

In contrast, the ML models employed in this study have demonstrated superior predictive performance. Unlike traditional formulas, these ML models are not constrained by predefined equations and can adapt to the complexities and non-linearities within the dataset. The models were trained and tuned using diverse features to capture a broader range of interactions between variables. Consequently, the ML models provided more accurate and generalized predictions across the entire dataset, outperforming the traditional formulas in most cases. The success of these models in this study suggests that they could serve as valuable tools for structural engineers seeking to predict bond strength with greater accuracy and reliability across diverse practical applications.

8. Conclusions

This study demonstrated the application of various ML models, including AdaBoost, CatBoost, Gradient Boosting, XGBoost, and Hist Gradient Boosting, in predicting the bond strength of reinforced concrete structures. Unlike traditional methods, ML models do not rely on explicit mathematical equations to predict rebar bond strength. Instead, these models are trained on experimental data, allowing them to capture complex patterns and relationships between input variables and bond strength. By learning from the data, ML models can provide accurate predictions without predefined equations. The employed models were thoroughly evaluated before and after hyperparameter tuning to assess their predictive capabilities.

- The results indicated that hyperparameter tuning significantly improved the performance of all models, particularly those that initially exhibited lower accuracy, such as AdaBoost and Hist Gradient Boosting.
- The analysis revealed that CatBoost, Gradient Boosting, and XGBoost consistently outperformed the other models, with XGBoost achieving the highest predictive accuracy after tuning. This was further corroborated by the Taylor diagrams, which illustrated the robustness of these models across training, testing, and combined datasets.
- The study also explored using a Voting Regressor to combine the strengths of multiple models. The findings showed that a Voting Regressor combining only the best-performing models (CatBoost, Gradient Boosting, and XGBoost) slightly improved predictive accuracy, demonstrating the value of model voting in enhancing prediction reliability.
- In addition to traditional feature importance analysis, SHAP values were employed to gain deeper insights into the impact of individual features on the model's predictions. The SHAP analysis highlighted that embedment length had a significant impact on predictions.
- The insights gained from this study underscore the importance of hyperparameter optimization and advanced interpretability techniques such as SHAP values in developing and evaluating machine learning models for structural engineering applications. The consistent identification of key features such as tensile strength, elastic modulus, and embedment length across different models and analyses reinforces their critical role in predicting bond strength, providing valuable guidance for future research and practical applications in this field.
- The findings demonstrate that while traditional predictive formulas can provide insights within specific experimental contexts, their limited generalizability highlights the need for more adaptable approaches, such as ML models, which have proven to deliver more accurate and reliable bond strength predictions across diverse scenarios.
- The user interface developed in this study enhances accessibility and practical application by allowing engineers to seamlessly implement and evaluate ML models for bond strength prediction in FRP-reinforced UHPC. By providing an interactive platform that supports customization of model parameters and real-time evaluation of model performance, the user interface bridges the gap between advanced ML techniques and their practical application in structural engineering. This tool empowers users to leverage state-of-the-art predictive models, thereby contributing to more accurate and efficient design and analysis processes in the field.

Author Contributions: Writing—review and editing, A.M. and M.B.; Conceptualization, A.M., M.B., and M.K.; Methodology, AM., M.B., and M.K.; Software, A.M.; Data curation, A.M., M.B.; Writing—review and editing, A.M., M.B., and M.K.; Investigation, A.M., M.B., and M.K.; Visualization, A.M. and M.B.; Supervision, M.K. All authors have read and agreed to the published version of the manuscript.

Funding: This research received no external funding.

Data Availability Statement: The research's complete dataset has been uploaded on GitHub, accessed on 5 October 2024 (https://github.com/AlirezaMahmoudian/GFRP_UHPC-Boosting-ML-models).

Conflicts of Interest: The authors declare no conflicts of interest.

References

1. El-Nemr, A.; Ahmed, E.A.; Barris, C.; Benmokrane, B. Bond-dependent coefficient of glass-and carbon-FRP bars in normal-and high-strength concretes. *Constr. Build. Mater.* **2016**, *113*, 77–89. [CrossRef]
2. Ahmed, E.A.; El-Sayed, A.K.; El-Salakawy, E.; Benmokrane, B. Bend strength of FRP stirrups: Comparison and evaluation of testing methods. *J. Compos. Constr.* **2010**, *14*, 3–10. [CrossRef]
3. ACI Committee 440. *440.1R-06: Guide for the Design and Construction of Concrete Reinforced with FRP Bars*; ACI: Farmington Hills, MI, USA, 2006.
4. Chin, W.J.; Park, Y.H.; Cho, J.-R.; Lee, J.-Y.; Yoon, Y.-S. Flexural behavior of a precast concrete deck connected with headed GFRP rebars and UHPC. *Materials* **2020**, *13*, 604. [CrossRef]
5. Kim, B.; Doh, J.-H.; Yi, C.-K.; Lee, J.-Y. Effects of structural fibers on bonding mechanism changes in interface between GFRP bar and concrete. *Compos. Part B Eng.* **2013**, *45*, 768–779. [CrossRef]
6. Yoo, D.-Y.; Banthia, N.; Yoon, Y.-S. Flexural behavior of ultra-high-performance fiber-reinforced concrete beams reinforced with GFRP and steel rebars. *Eng. Struct.* **2016**, *111*, 246–262. [CrossRef]
7. Saleh, N.; Ashour, A.; Lam, D.; Sheehan, T. Experimental investigation of bond behaviour of two common GFRP bar types in high–Strength concrete. *Constr. Build. Mater.* **2019**, *201*, 610–622. [CrossRef]
8. Yoo, S.-J.; Hong, S.-H.; Yoon, Y.-S. Bonding behavior and prediction of helically ribbed CFRP bar embedded in ultra high-performance concrete (UHPC). *Case Stud. Constr. Mater.* **2023**, *19*, e02253. [CrossRef]
9. Taha, A.; Alnahhal, W. Bond durability and service life prediction of BFRP bars to steel FRC under aggressive environmental conditions. *Compos. Struct.* **2021**, *269*, 114034. [CrossRef]
10. Hassan, M.; Benmokrane, B.; ElSafty, A.; Fam, A. Bond durability of basalt-fiber-reinforced-polymer (BFRP) bars embedded in concrete in aggressive environments. *Compos. Part B Eng.* **2016**, *106*, 262–272. [CrossRef]
11. Michaud, D.; Fam, A. Development length of small-diameter basalt FRP bars in normal-and high-strength concrete. *J. Compos. Constr.* **2021**, *25*, 04020086. [CrossRef]
12. Jeddian, S.; Ghazi, M.; Sarafraz, M.E. Experimental study on the seismic performance of GFRP reinforced concrete columns actively confined by AFRP strips. *Structures.* **2024**, *62*, 106248. [CrossRef]
13. Liang, K.; Chen, L.; Shan, Z.; Su, R. Experimental and theoretical study on bond behavior of helically wound FRP bars with different rib geometry embedded in ultra-high-performance concrete. *Eng. Struct.* **2023**, *281*, 115769. [CrossRef]
14. Yang, K.; Wu, Z.; Zheng, K.; Shi, J. Design and flexural behavior of steel fiber-reinforced concrete beams with regular oriented fibers and GFRP bars. *Eng. Struct.* **2024**, *309*, 118073. [CrossRef]
15. Ding, Y.; Ning, X.; Zhang, Y.; Pacheco-Torgal, F.; Aguiar, J. Fibres for enhancing of the bond capacity between GFRP rebar and concrete. *Constr. Build. Mater.* **2014**, *51*, 303–312. [CrossRef]
16. Jamshaid, H.; Mishra, R. A green material from rock: Basalt fiber–a review. *J. Text. Inst.* **2016**, *107*, 923–937. [CrossRef]
17. Yoshitake, I.; Hasegawa, H.; Shimose, K. Monotonic and cyclic loading tests of reinforced concrete beam strengthened with bond-improved carbon fiber reinforced polymer (CFRP) rods of ultra-high modulus. *Eng. Struct.* **2020**, *206*, 110175. [CrossRef]
18. Jia, L.; Fang, Z.; Hu, R.; Pilakoutas, K.; Huang, Z. Fatigue behavior of UHPC beams prestressed with external CFRP tendons. *J. Compos. Constr.* **2022**, *26*, 04022066. [CrossRef]
19. Fang, Y.; Fang, Z.; Feng, L.; Xiang, Y.; Zhou, X. Bond behavior of an ultra-high performance concrete-filled anchorage for carbon fiber-reinforced polymer tendons under static and impact loads. *Eng. Struct.* **2023**, *274*, 115128. [CrossRef]
20. Pan, R.; Zou, J.; Liao, P.; Dong, S.; Deng, J. Effects of fiber content and concrete cover on the local bond behavior of helically ribbed GFRP bar and UHPC. *J. Build. Eng.* **2023**, *80*, 107939. [CrossRef]
21. Shaikh, F.U.A.; Luhar, S.; Arel, H.Ş.; Luhar, I. Performance evaluation of Ultrahigh performance fibre reinforced concrete–A review. *Constr. Build. Mater.* **2019**, *232*, 117152. [CrossRef]
22. Tan, H.; Hou, Z.; Li, Y.; Xu, X. A flexural ductility model for UHPC beams reinforced with FRP bars. *Structures.* **2022**, *45*, 773–786. [CrossRef]
23. Ke, L.; Liang, L.; Feng, Z.; Li, C.; Zhou, J.; Li, Y. Bond performance of CFRP bars embedded in UHPFRC incorporating orientation and content of steel fibers. *J. Build. Eng.* **2023**, *73*, 106827. [CrossRef]

24. Zhu, H.; He, Y.; Cai, G.; Cheng, S.; Zhang, Y.; Larbi, A.S. Bond performance of carbon fiber reinforced polymer rebars in ultra-high-performance concrete. *Constr. Build. Mater.* **2023**, *387*, 131646. [CrossRef]
25. Wu, C.; Ma, G.; Hwang, H.-J. Bond performance of spliced GFRP bars in pre-damaged concrete beams retrofitted with CFRP and UHPC. *Eng. Struct.* **2023**, *292*, 116523. [CrossRef]
26. *CSA S806-2012*; Design and Construction of Building Structures with Fibre Reinforced Polymers. Canadian Standard Association: Mississauga, ON, Canada, 2012.
27. Mahaini, Z.; Abed, F.; Alhoubi, Y.; Elnassar, Z. Experimental and numerical study of the flexural response of Ultra High Performance Concrete (UHPC) beams reinforced with GFRP. *Compos. Struct.* **2023**, *315*, 117017. [CrossRef]
28. Mahmoudian, A.; Bypour, M.; Kontoni, D.-P.N. Tree-based machine learning models for predicting the bond strength in reinforced recycled aggregate concrete. *Asian J. Civ. Eng.* **2024**, 1–26. [CrossRef]
29. Bypour, M.; Mahmoudian, A.; Tajik, N.; Taleshi, M.M.; Mirghaderi, S.R.; Yekrangnia, M. Shear capacity assessment of perforated steel plate shear wall based on the combination of verified finite element analysis, machine learning, and gene expression programming. *Asian J. Civ. Eng.* **2024**, *25*, 5317–5333. [CrossRef]
30. Feng, D.-C.; Liu, Z.-T.; Wang, X.-D.; Chen, Y.; Chang, J.-Q.; Wei, D.-F.; Jiang, Z.-M. Machine learning-based compressive strength prediction for concrete: An adaptive boosting approach. *Constr. Build. Mater.* **2020**, *230*, 117000. [CrossRef]
31. Kaloop, M.R.; Kumar, D.; Samui, P.; Hu, J.W.; Kim, D. Compressive strength prediction of high-performance concrete using gradient tree boosting machine. *Constr. Build. Mater.* **2020**, *264*, 120198. [CrossRef]
32. Zhu, F.; Wu, X.; Lu, Y.; Huang, J. Strength estimation and feature interaction of carbon nanotubes-modified concrete using artificial intelligence-based boosting ensembles. *Buildings* **2024**, *14*, 134. [CrossRef]
33. Hossain, K.; Ametrano, D.; Lachemi, M. Bond strength of GFRP bars in ultra-high strength concrete using RILEM beam tests. *J. Build. Eng.* **2017**, *10*, 69–79. [CrossRef]
34. Ahmad, F.S.; Foret, G.; Le Roy, R. Bond between carbon fibre-reinforced polymer (CFRP) bars and ultra high performance fibre reinforced concrete (UHPFRC): Experimental study. *Constr. Build. Mater.* **2011**, *25*, 479–485. [CrossRef]
35. Sayed-Ahmed, M.; Sennah, K. Str-894: Bond strength of ribbed-surface high-modulus glass FRP bars embedded into unconfined UHPFRC. In Proceedings of the Canadian Society for Civil Engineering Annual Conference 2016, Resilient Infrastructure, London, UK, 1–4 June 2016.
36. Michaud, D.; Fam, A.; Dagenais, M.-A. Development length of sand-coated GFRP bars embedded in Ultra-High performance concrete with very small cover. *Constr. Build. Mater.* **2021**, *270*, 121384. [CrossRef]
37. Yoo, D.-Y.; Yoon, Y.-S. Bond behavior of GFRP and steel bars in ultra-high-performance fiber-reinforced concrete. *Adv. Compos. Mater.* **2017**, *26*, 493–510. [CrossRef]
38. Yoo, D.-Y.; Kwon, K.-Y.; Park, J.-J.; Yoon, Y.-S. Local bond-slip response of GFRP rebar in ultra-high-performance fiber-reinforced concrete. *Compos. Struct.* **2015**, *120*, 53–64. [CrossRef]
39. Yoo, S.-J.; Hong, S.-H.; Yoo, D.-Y.; Yoon, Y.-S. Flexural bond behavior and development length of ribbed CFRP bars in UHPFRC. *Cem. Concr. Compos.* **2024**, *146*, 105403. [CrossRef]
40. Hu, X.; Xue, W.; Xue, W. Bond properties of GFRP rebars in UHPC under different types of test. *Eng. Struct.* **2024**, *314*, 118319. [CrossRef]
41. Qasem, A.; Sallam, Y.S.; Eldien, H.H.; Ahangarn, B.H. Bond-slip behavior between ultra-high-performance concrete and carbon fiber reinforced polymer bars using a pull-out test and numerical modelling. *Constr. Build. Mater.* **2020**, *260*, 119857. [CrossRef]
42. Eltantawi, I.; Alnahhal, W.; El Refai, A.; Younis, A.; Alnuaimi, N.; Kahraman, R. Bond performance of tensile lap-spliced basalt-FRP reinforcement in high-strength concrete beams. *Compos. Struct.* **2022**, *281*, 114987. [CrossRef]
43. ACI Committee 440. *Guide Test Methods for FiberReinforced Polymer (FRP) Composites for Reinforcing or Strengthening Concrete and Masonry Structures*; ACI: Farmington Hills, MI, USA, 2012.
44. Tong, D.; Chi, Y.; Huang, L.; Zeng, Y.; Yu, M.; Xu, L. Bond performance and physically explicable mathematical model of helically wound GFRP bar embedded in UHPC. *J. Build. Eng.* **2023**, *69*, 106322. [CrossRef]
45. Ahmed, M.S.; Sennah, K. Pullout strength of sand-coated GFRP bars embedded in ultra-high performance fiber reinforced concrete. In Proceedings of the CSCE 2014 4th International Structural Speciality, Halifax, NS, Canada, 28–31 May 2014.
46. Esteghamati, M.Z.; Gernay, T.; Banerji, S. Evaluating fire resistance of timber columns using explainable machine learning models. *Eng. Struct.* **2023**, *296*, 116910. [CrossRef]
47. Mahmoudian, A.; Tajik, N.; Taleshi, M.M.; Shakiba, M.; Yekrangnia, M. Ensemble machine learning-based approach with genetic algorithm optimization for predicting bond strength and failure mode in concrete-GFRP mat anchorage interface. *Structures* **2023**, *57*, 105173. [CrossRef]
48. Tajik, N.; Mahmoudian, A.; Taleshi, M.M.; Yekrangnia, M. Explainable XGBoost machine learning model for prediction of ultimate load and free end slip of GFRP rod glued-in timber joints through a pull-out test under various harsh environmental conditions. *Asian J. Civ. Eng.* **2024**, *25*, 141–157. [CrossRef]
49. Shams, M.Y.; Tarek, Z.; Elshewey, A.M.; Hany, M.; Darwish, A.; Hassanien, A.E. A machine learning-based model for predicting temperature under the effects of climate change. In *The Power of Data: Driving Climate Change with Data Science and Artificial Intelligence Innovations*; Studies in Big Data; Springer: Berlin/Heidelberg, Germany, 2023; pp. 61–81.

50. Alhakeem, Z.M.; Jebur, Y.M.; Henedy, S.N.; Imran, H.; Bernardo, L.F.A.; Hussein, H.M. Prediction of ecofriendly concrete compressive strength using gradient boosting regression tree combined with GridSearchCV hyperparameter-optimization techniques. *Materials* **2022**, *15*, 7432. [CrossRef]
51. Lee, J.-Y.; Kim, T.-Y.; Kim, T.-J.; Yi, C.-K.; Park, J.-S.; You, Y.-C.; Park, Y.-H. Interfacial bond strength of glass fiber reinforced polymer bars in high-strength concrete. *Compos. Part B Eng.* **2008**, *39*, 258–270. [CrossRef]

computation

MDPI

Article

Numerical Homogenization Method Applied to Evaluate Effective Converse Flexoelectric Coefficients

Khader M. Hamdia

Independent Researcher, 31535 Hannover, Germany; khaderhamdia@gmail.com

Abstract: This paper presents a numerical homogenization method for estimating the effective converse flexoelectric coefficients. A 2D model made of two-phase composite is developed at the microscale in consideration of a representative volume element that includes a continuous flexoelectric fiber embedded in a pure elastic matrix. In the implementation, the constitutive equations are derived from the electromechanical enthalpy accounting for higher-order coupling terms. Electric boundary conditions associated with an inhomogeneous electric field are imposed, allowing the approximation of the generated mechanical strains and stresses. Accordingly, the numerical simulations yield the overall equivalent converse flexoelectricity tensor for the longitudinal, transversal, and shear couplings. The results showed that the composite undergoes an obvious straining, which creates actuation due to the converse effect. The components of the homogenized longitudinal and transverse coefficients were found to be dependent on the volume fraction and elastic properties of the constituents.

Keywords: numerical homogenization; electromechanical coupling; converse flexoelectricity; flexoelectric nanocomposites

1. Introduction

Flexoelectricity is a physical property related to the break of the symmetry inversion in dielectrics. In this case, the electromechanical coupling is defined by the relationship between the displacement of ions (i.e., polarization) and the mechanical deformation. The energy conversion mechanism is, at best, notable at small length scales. With recent developments in nanotechnology, increasing interest has been shown in studying the modification in the electromechanical behavior of flexoelectric nanocomposites that provides the possibility to produce considerably larger flexoelectric outcomes [1,2]. The concept of direct flexoelectricity was introduced by Kogan in the 1960s, who quantified the effect in solid crystals by a coefficient that couples the strain gradient and the induced electric polarization [3]. In this direct form, an electrical polarization is induced in the presence of an inhomogeneous strain. Later, Tagantsev [4] extended the phenomenological flexoelectricity theory of Kogan by accounting for surface and bulk contributions [4]. Further experimental investigations have confirmed the theory of Tagantsev for the role of electric susceptibility in improving the flexoelectric effect [5]. In addition, several theoretically based studies have been devoted to better understand the fundamentals of electromechanical coupling [2,6,7].

In contrast, the converse effect of flexoelectricity causes a deformation when applying an electric field. The direct and converse coefficients are widely used to measure flexoelectric performance. Still, their sign and relative magnitude are contradictory in the relevant literature. Flexoelectric effects have been experimentally reported in the range of $10^{-9} - 10^{-6}$ C/m for certain ferroelectrics and perovskite oxides [8]. To calculate the

 https://doi.org/10.3390/computation13020048

converse coefficients, the common setup includes measuring the bending in a capacitor after applying a voltage or applying a voltage across a truncated pyramid; see [9–11] among others. Analogously, a large effective piezoelectric response has been produced in nonpiezoelectric materials, and the converse flexoelectricity showed non-negligible effects in thin films [12]. The equivalent piezoelectric effect has been shown to increase with the converse flexoelectricity in permittivity gradient composites [13]. The mechanism has been investigated in an asymmetric structure around 90° domain walls at which large polarization and strain gradients are clearly concentrated [14]. An analysis of atomic-scale electron microscopy imaging data has been integrated with phenomenological phase field modeling to study the influence of flexocoupling on the global vortex structure [15]. Moreover, flexoelectric coefficients were extracted for a homogeneous domain from simulations based on ab initio lattice dynamics [16] and first-principles calculations [17]. An explicit solution was presented relying on a micro-hole model. The formulations were based on higher-order governing equations that take into account the strain gradient elasticity, the direct flexoelectricity, and the converse flexoelectricity [18]. Approaches along these lines were implemented to exact solutions for the flexoelectric response in [19].

In two- (or multiple-)phase composites, an active material which has the flexoelectric properties can increase the efficiency of energy conversion in the zones of high strain gradients. A significant enhancement of polarization could be attributed to the flexoelectric effect [20]. Nevertheless, it has been shown that it is possible to produce piezoelectric nanocomposites without the use of piezoelectric materials [21]. In this regard, elastic heterogeneity impacts the effective piezo- and flexoelectric responses as well. Analytical and computational homogenization have provided insight into the coupling at the microscale to determine the average macroscopic properties where a heterogeneous domain is transformed into a small body of a homogeneous one. Worth mentioning here are the documented works for piezoelectric, electrets, and electrostrictive composites in [22–25]. The homogenized flexoelectric, dielectric, and elastic response was theoretically studied by [26]. In the mean time, an analytical micromechanic approach was exploited to calculate the effective flexoelectric coefficients in a multifunctional fiber composite [27]. Asymptotic homogenization was applied to flexoelectric rods [28] and flexoelectric composite plates with a periodically varying thickness [29]. Considering the flexoelectric effect, the effective elastic piezoelectric constants were examined via analytical-based micromechanical models and finite element models [30–32].

The numerical implementation of computational homogenization commonly used a representative volume element (RVE) to approximate the kinematic and constitutive equations in equilibrium. In [33], an apparent flexoelectricity response was concluded in heterogeneous piezoelectricity by applying a computational approach on the finite element method. As the flexoelectricity of the constituent materials was not considered, the effective flexoelectric tensor was approximated in different geometries. Similarly, an inclusion-based piezoelectric planar composite was examined, and the effective properties were predicted using a variational formulation [34]. The effective material properties and flexoelectric coefficients were evaluated using a 3D RVE with periodic boundary conditions in [35]. The numerical simulation was implemented on a microstructure composed of an elastic matrix and cylindrical flexoelectric filler. Ref. [36] has solved the higher-order electromechanical coupling using a high-order approximation space constrained to periodicity conditions. Special attention was given to the homogenization of architected metamaterials with apparent piezoelectricity. It is worth noting that the above-mentioned works have focused on the direct effects of flexoelectricity. With regard to the converse effect, a numerical study was provided in [37] which revealed an enhanced overall flexoelectric response in periodic piezoelectric composites. The apparent converse coefficients were found to be in the same order as the direct coefficients in the local constituents. Indeed, to the author's best knowledge, the contributions in this direction are scarce. A thorough and robust numerical technique is at high demand, as it can provide a deeper understanding of the direct and converse flexoelectricity.

The main objective of this study is to evaluate the effect of converse flexoelectricity in two-phase composites, namely elastic and flexoelectric. In particular, a numerical homogenization method is presented, offering significant advantages in predicting the effective converse flexoelectric coefficients. The electromechanical coupling terms associated with the strain, the electric field gradient, and the strain gradient elasticity are derived from the electrical enthalpy. Non-Uniform Rational B-spline (NURBS)-based isogeometric analysis (IGA) is used to address the problem leveraging its higher-order continuity. The study relies on numerical simulations conducted on a two-dimensional model subjected to inhomogeneous electric field. The solution involves computing the distribution of the electrical field, as well as the generated stress and strain fields. Then, the overall performance of the heterogeneous system is obtained by calculating the equivalent coefficients. The converse flexoelectricity tensor is evaluated at varying volume fractions including comparison with respect to the ratio of the elasticity between the two constituents. The paper is organized as follows. Section 2 introduces the mathematical formulation of the continuum flexoelectric boundary value problem, and presents how NURBS-based isogeometric analysis (IGA) is used to approximate the solution of the electromechanically coupled system. Section 3 describes the setup of the problem, including an in-depth discussion of the results of computational experiments. Afterwards, the summary and conclusions are shown in Section 4.

2. Flexoelectricity Modeling Formulation

Direct and converse flexoelectricity are characterized by fourth-order tensors that couple polarization and strain gradient in the former, and strain and polarization gradient in the latter. Considering flexoelectricity theory, the internal energy includes additional terms related to the electromechanical coupling. The Gibbs-free (electromechanical enthalpy) bulk energy for a linear flexoelectric solid can be written as

$$\begin{aligned}\mathcal{H}\left(\varepsilon_{ij}, E_i, \varepsilon_{jk,l}, E_{k,l}\right) = & \frac{1}{2}\mathbb{C}_{ijkl}\varepsilon_{ij}\varepsilon_{kl} - \frac{1}{2}\kappa_{ij}E_iE_j - e_{ikl}E_i\varepsilon_{kl} \\ & + d_{ijkl}E_i\varepsilon_{jk,l} + f_{ijkl}E_{i,j}\varepsilon_{kl} + \frac{1}{2}g_{ijklmn}\varepsilon_{ij,k}\varepsilon_{lm,n}\end{aligned} \quad (1)$$

where ε_{ij} is the strain tensor and $E_i = -\phi_{,i}$ is the vector of the electric field, with ϕ being the electric potential. The comma notation $(,)$ in the subscript denotes the first derivative with respect to one spatial variable. The first two terms represent elastic and electrostatic potentials, respectively, where $\mathbb{C}_{ijkl}$ is the fourth-order tensor of elastic moduli and κ_{ij} indicates the second-order tensor of dielectric permittivity. The contribution of piezoelectricity is represented by the third-order piezoelectric tensor, e_{ikl}, which couples the strain and the electric field. Likewise, the flexoelectric effect is accounted for by fourth-order flexocoupling tensors in the fourth and fifth terms in the above expression. Here, d_{ijkl} and f_{ijkl} represent the direct and converse tensors, respectively. For cubic symmetry materials, each tensor has only three nonzero independent components: longitudinal, transverse, and shear coefficients [38]. The particular focus hereinafter is on the converse constants, which correspond to the ratio of the polarization gradient to the produced strain. In addition, this study considers the nonlocal elastic energy associated with the strain gradient elasticity. To ensure positive strain energy, a relatively small value is assumed for a length scale parameter (l) according to the simplified model of strain gradient elasticity theory, such that the sixth-order elastic tensor $g_{ijklmn} = l^2\mathbb{C}_{jkmn}$ [39]. Further higher-order coupling terms are neglected for the sake of simplicity.

From the above-mentioned enthalpy, the constitutive relations for the electric displacement ($\mathbf{D}$), the stress tensor (σ), and the higher-order tensors (hyper) of stress ($\mathbf{S}$) and electric displacement ($\mathbf{P}$, known as the electric quadrupole) are obtained as follows:

$$D_i = e_{ikl}\varepsilon_{kl} + \kappa_{ij}E_j - d_{ijkl}\varepsilon_{jk,l} \tag{2a}$$

$$\sigma_{ij} = \mathbb{C}_{ijkl}\varepsilon_{kl} - e_{ikl}E_i + f_{ijkl}E_{i,j} \tag{2b}$$

$$S_{ijk} = d_{lijk}E_l + g_{ijklmn}\varepsilon_{lm,n} \tag{2c}$$

$$P_{ij} = f_{iklj}\varepsilon_{kl} \tag{2d}$$

When solving the weak form, it is required to solve fourth-order partial differential equations (PDEs), which necessitates C^1 continuity. This study employs the isogeometric analysis (IGA) approach for the discretization, because it includes higher-order smooth basis functions. For this purpose, the formulation presented in [40,41] is adopted based on Non-Uniform Rational B-spline (NURBS) basis functions, $N_{(i,p)}(\xi)$. The geometry is discretized by control points. In doing so, mechanical and electrical degrees of freedom are assigned for each control point. To approximate the solution of field variables, the nodal values are multiplied with their corresponding basis functions. The mechanical displacement ($\mathbf{u}$) and electric potential (ϕ) fields with the corresponding variations are approximated by

$$\mathbf{u} = \mathbf{N}_{\mathrm{u}}{}^{\mathrm{T}}\mathbf{u}^{\mathrm{e}} \tag{3a}$$

$$\phi = \mathbf{N}_{\phi}{}^{\mathrm{T}}\boldsymbol{\phi}^{\mathrm{e}} \tag{3b}$$

$$\partial_j u = \partial_j(\mathbf{N}_{\mathrm{u}})^{\mathrm{T}}\mathbf{u}^{\mathrm{e}} = (\mathbf{B}_{\mathrm{u}})^{\mathrm{T}}\mathbf{u}^{\mathrm{e}} = \boldsymbol{\varepsilon} \tag{3c}$$

$$\partial_j\theta = \partial_j(\mathbf{N}_{\phi})^{\mathrm{T}}\boldsymbol{\phi}^{\mathrm{e}} = (\mathbf{B}_{\phi})^{\mathrm{T}}\boldsymbol{\phi}^{\mathrm{e}} = -\mathbf{E} \tag{3d}$$

$$\partial_j\partial_k u = \partial_j\partial_k(\mathbf{N}_{\mathrm{u}})^{\mathrm{T}}\mathbf{u}^{\mathrm{e}} = (\mathbf{H}_{\mathrm{u}})^{\mathrm{T}}\mathbf{u}^{\mathrm{e}} = \nabla\boldsymbol{\varepsilon} \tag{3e}$$

$$\partial_j\partial_k\phi = \partial_j\partial_k(\mathbf{N}_{\phi})^{\mathrm{T}}\boldsymbol{\phi}^{\mathrm{e}} = (\mathbf{H}_{\mathrm{s}})^{\mathrm{T}}\boldsymbol{\phi}^{\mathrm{e}} = -\nabla\mathbf{E} \tag{3f}$$

where the superscript e denotes nodal parameters at the control points. The gradient operators $\mathbf{B}_{\mathrm{u}}$ and $\mathbf{B}_{\phi}$ are the spatial derivatives of the basis functions $\mathbf{N}_{\mathrm{u}}$ and $\mathbf{N}_{\phi}$, respectively, while $\mathbf{H}_{\mathrm{u}}$ and $\mathbf{H}_{\mathrm{s}}$ are their second-order (Hessian) derivatives. Note that the NURBS basis functions are continuously differentiable up to $p-1$ a polynomial order of p [42].

The boundary conditions can be imposed as mechanical tractions, $\bar{\mathbf{t}}$, or surface charge density, $\boldsymbol{\varpi}$, at the boundaries Γ_t and Γ_D. By integration over the domain Ω, the discrete representation of Equation (1), after some algebra, reads [41]

$$\begin{aligned} H = {} & \frac{1}{2}\sum_{\mathrm{e}}\mathbf{u}^{\mathrm{T}}\left(\int_{\Omega_{\mathrm{e}}}\mathbf{B}_{\mathrm{u}}\mathbb{C}(\mathbf{B}_{\mathrm{u}})^{\mathrm{T}}d\Omega_{\mathrm{e}}\right)\mathbf{u} - \frac{1}{2}\sum_{\mathrm{e}}\left(\int_{\Omega_{\mathrm{e}}}\mathbf{B}_{\phi}\kappa(\mathbf{B}_{\phi})^{\mathrm{T}}d\Omega_{\mathrm{e}}\right)\phi\phi + \sum_{\mathrm{e}}\mathbf{u}^{\mathrm{T}}\left(\int_{\Omega_{\mathrm{e}}}\mathbf{B}_{\mathrm{u}}e(\mathbf{B}_{\phi})^{\mathrm{T}}d\Omega_{\mathrm{e}}\right)\phi \\ & + \sum_{\mathrm{e}}\mathbf{u}^{\mathrm{T}}\left(\int_{\Omega_{\mathrm{e}}}\mathbf{H}_{\mathrm{u}}d^{\mathrm{T}}(\mathbf{B}_{\phi})^{\mathrm{T}}d\Omega_{\mathrm{e}}\right)\phi + \sum_{\mathrm{e}}\mathbf{u}^{\mathrm{T}}\left(\int_{\Omega_{\mathrm{e}}}(\mathbf{B}_{\mathrm{u}})^{\mathrm{T}}f(\mathbf{H}_{\phi})^{\mathrm{T}}d\Omega_{\mathrm{e}}\right)\phi + \frac{1}{2}\sum_{\mathrm{e}}\mathbf{u}^{\mathrm{T}}\left(\int_{\Omega_{\mathrm{e}}}\mathbf{H}_{\mathrm{s}}\mathbf{g}(\mathbf{H}_{\mathrm{s}})^{\mathrm{T}}d\Omega_{\mathrm{e}}\right)\mathbf{u} \\ & - \sum_{\mathrm{e}}\left(\int_{\Gamma_{t_{\mathrm{e}}}}\mathbf{N}_{\mathrm{u}}^{\mathrm{T}}\mathbf{t}_{\Gamma}ds\right)\mathbf{u} + \sum_{\mathrm{e}}\left(\int_{\Gamma_{D_{\mathrm{e}}}}\mathbf{N}_{\phi}^{\mathrm{T}}\varpi ds\right)\phi \end{aligned} \tag{4}$$

The corresponding mechanical and electrostatic equilibrium are obtained for $\mathbf{u}$ and ϕ, respectively. Hence, the global stiffness coupled matrix is assembled from the mechanical, electromechanical, and electrical elemental stiffness matrices [43].

3. Numerical Implementation and Results

This study investigates a two-phase composite constituted by flexo (active phase) and elastic (passive phase) materials under the assumption of plane strain conditions. The former experiences pure flexoelectricity (i.e, non-piezoelectric with $e_{ij} = 0$, $\kappa_{ij} = 12.48 \times 10^{-9}$ C/Vm and $f_{ijkl} = 1.0 \times 10^{-6}$ C/m), whereas the latter is a plain elastic material. This can provide a clear distinction between their individual physics. The geometry of the reference problem is sketched in Figure 1, which takes the shape of a square RVE with dimensions $L_{x_1} = L_{x_2} = 100 \times 10^{-6}$ m. A two-dimensional configuration with longitudinally oriented active fibers is being considered. This configuration enables the study of the converse effect in both the longitudinal and transverse directions.

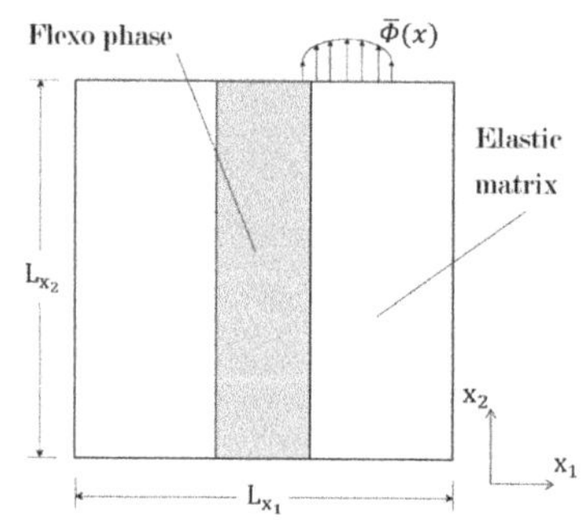

Figure 1. Schematic representation of the RVE.

In the analysis, a transversely isotropic elasticity is assumed, incorporating softer passive material that can yield enhanced deflection. The elastic properties for the matrix are taken as $\mathbb{C}_{11}$ = 60 GPa, $\mathbb{C}_{12}$ = 32.3 GPa, and $\mathbb{C}_{44}$ = 13.8 GPa; while for the flexo phase, they are $\mathbb{C}_{11}$ = 120 GPa, $\mathbb{C}_{12}$ = 40 GPa, and $\mathbb{C}_{44}$ = 40 GPa. However, we will change these properties and examine their effects on the effective coefficients at a varied volume fraction. A no-slippage condition is accounted for, ensuring continuity in displacement, traction, and electrical potential along the interfaces.

While the effective dielectric properties can be estimated using a uniform electric field, the homogenized piezoelastic (piezoelectric and elastic) constants can be obtained under boundary conditions of uniform strain. Differently, the direct flexoelectric constants can be determined by imposing a strain gradient. However, the scope of this study is to evaluate the converse flexoelectricity. Boundary conditions that create a polarization gradient are considered for this purpose. An electric field gradient is generated independently, ensuring that all other kinematic conditions remain at zero, except for the component under investigation. Note in this case that $\overline{\mathbf{F}} = \overline{\mathbf{P}} : [\bar{\varepsilon}]^{-1}$. In doing so, the RVE is subjected to a periodically fluctuating electric potential, ϕ, which will result in non-homogeneous electric fields. Quadratic electric boundary conditions for the system are applied at the surfaces/boundaries as follows:

$$\phi|_{\mathrm{x}=0,\mathrm{L}} = \phi(\mathrm{x}) = V_0 - V_0\left(\frac{2\mathrm{x}}{\mathrm{L}} - 1\right)^2 \quad (5)$$

where V_0 is the amplitudes of the electric potential.

For the numerical implementation, the model with the established geometry is meshed by quadratic B-spline elements. The control points are utilized to discretize the geometry and encompass the degrees of freedom (DoF). Two DoF pertaining to mechanical displacement and one for the electric potential are assigned. Mesh sensitivity is evaluated using different sizes associated with a quadratic shape function in all simulations. It has been determined that discretization of 40 × 40 elements is sufficient to achieve accurate results, revealing only a negligible impact on the numerical approximation. Macroscopic quantities are obtained via the volume averaging operator $\overline{(\cdot)} = \langle\cdot\rangle := {}^{1}/_{V} \int_{\Omega}(\cdot)dV$.

Throughout all simulations, we consider the converse flexoelectricity effect under predefined electrical loading. An electric field gradient with respect to x_2, $(\nabla E_{22} = E_{2,2})$, can be obtained by applying a non-uniform electric potential on the left and right edges of the RVE in accordance to Equation (5). In similar way, ∇E_{11} is generated from a non-uniform voltage on the top and bottom surfaces. Figure 2 illustrates the distribution of the axial stresses σ_{22} within an RVE subjected to an electric gradient ∇E_{22}. The microstructure is loaded with flexo material ratio of 20% by volume and the amplitude V_0 is 100 V. Negative and positive concentrations of electric quadrupoles are induced in the x_1 and x_2 directions, respectively, within the flexo phase due to the electric gradient. The contours of stress fields in Figure 2a exhibit an axis of symmetry along x_2-axis attributed to the symmetrical microstructure along this axis. Furthermore, the applied electrical loading primarily induces negative stresses within the inclusion and positive stresses in its surrounding matrix, because of perfect bonding of the two materials and the absence of boundary constraints. Consequently, the flexo phase contracted while the matrix expanded due to stresses. The direction of the electric potential determines the sign of the straining. It can be revealed that the generated stresses are a clear indication of actuation as a result of the converse flexoelectric effect. The stress profiles across the mid height calculated at different loadings are depicted in Figure 2b. The plots confirm that the axial stress developed through the flexo phase is considerably stronger. This is attributed to the flexoelectric coupling. It is also observed that with an increase in the magnitude of the applied voltage, a proportionate increase in the actuation is produced.

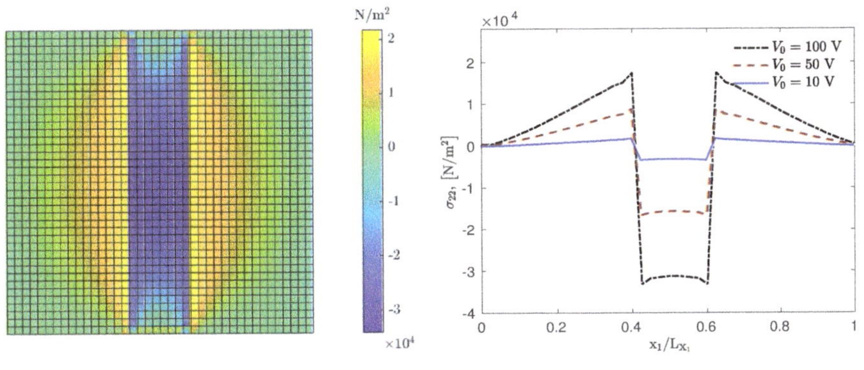

(**a**) (**b**)

Figure 2. Distribution of the axial stress σ_{22} due to an electric gradient ∇E_{22}: (**a**) contour of stress field scaled by 2 corresponding to V_0 = 100 V, and (**b**) variation of σ_{22} at mid-height for different electrical potentials.

Table 1 presents the homogenized converse coefficients of the longitudinal, transversal and shear couplings in a system including flexo phase comprising a volume fraction of 20%. The second row of the table includes the percentages of the equivalent coefficients relative to the flexo phase of $f_{ijkl} = 1.0 \times 10^{-6}$ C/m. The effective converse shear coefficient $\overline{f}_{1212}$ exhibits the largest value of 0.94 × 10^{-6} C/m, leading to a remarkable contribution to the shear strain. The components $\overline{f}_{2222}$ and $\overline{f}_{2211}$ are 0.49 × 10^{-6} C/m and 0.51 × 10^{-6} C/m, respectively. Interestingly, $\overline{f}_{2222}$ and $\overline{f}_{2211}$ are proportional to in-plane actuation, while $\overline{f}_{1212}$ is proportional to out-of-plane actuation. When reinforcing an elastic matrix with only 20% by volume, the overall coefficients can be improved by up to 50% of the flexo material. This clearly demonstrates the significant impact of converse flexoelectricity on electromechanical coupling.

Table 1. Effective converse flexoelectric coefficients and their ratios to the flexo phase of $f_{ijkl} = 1.0 \times 10^{-6}$ C/m.

	$\overline{f}_{ijkl}$ [$\times 10^{-6}$ C/m]		
	$\overline{f}_{2222}$	$\overline{f}_{2211}$	$\overline{f}_{1212}$
Coefficient	0.49	0.51	0.94
Relative percentage	49%	51%	94%

Additionally, the evolution of the components of the converse flexoelectric coefficients with respect to the volume fraction is studied and plotted in Figure 3. It also demonstrates the impact of material elasticity by considering different ratios of the elasticity modulus of the matrix relative to the inclusion, $\rho = \mathbb{C}_{11elastic}/\mathbb{C}_{11flexo}$. From the figure, it can be observed that the components $\overline{f}_{2222}$ and $\overline{f}_{2211}$, which couple the polarization gradient and strain in the axial and transverse directions, are characterized by an obvious increasing trend. The reason for these improvements is that the inclusion is considered with electromechanical properties while the matrix is purely elastic material. An upward shift of can be noticed as the ratio ρ increases from 0.5 to 2.0 implying a stronger converse effect. The change in $\overline{f}_{1212}$, is almost marginal over the entire range resulting in a relatively modest improvement, particularly at microstructure filled with $\geqslant 10\%$ volume fractions.

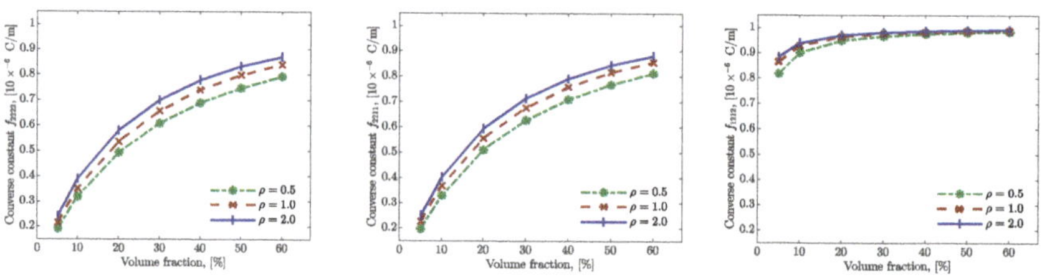

(a) (b) (c)

Figure 3. Variation of the effective converse flexoelectric coefficients versus the volume fraction: (**a**) f_{2222}, (**b**) f_{2211}, and (**c**) f_{1212}. $\rho = \mathbb{C}_{11elastic}/\mathbb{C}_{11flexo}$.

4. Summary

Flexoelectricity is a complex electromechanical coupling phenomenon that is challenging to fully comprehend and precisely quantify. The system is characterized by fourth-order partial differential equations, requiring C^1 continuity. The energy conversion relies on higher-order coupling terms that relate the electric polarization and mechanical strain, along with their corresponding gradients. The mechanism can stand for direct or converse effects concerning to the induced energy terms. The impact of strain and electric field gradients is evident in two-phase materials, because the high gradients can contribute more to the total energy.

In this study, numerical homogenization was employed to evaluate the overall effective converse coefficients in composites. The continuum flexoelectric boundary value problem was formulated and solved numerically accounting for the electrical enthalpy. The constitutive relationships are defined as the work conjugates of strain, electric field, and their gradients. Isogeometric analysis with Non-Uniform Rational B-spline (NURBS) was considered in solving the governing equations and approximating the solution. Particular attention was given to a heterogeneous composite system formed by unidirectional continuous fibers aligned in an elastic matrix. The fibers are non-piezoelectric and are assumed to feature purely flexoelectric behavior with a fully bonded interface with the surrounding matrix, involving continuity in both displacement and electric potential. For this purpose, a standard continuum model was adopted employing a 2D representative volume element at microstructure. The geometry was meshed with 40 × 40 B-Spline elements. Each control point was defined with two types of DoFs, namely conventional displacement and electric potential. The study examined a pre-designed electrical loading as the RVE was subjected to fluctuated electric potential, resulting in non-homogeneous electric fields. Subsequently, the homogenized values were obtained by integrating over the volume.

The simulations revealed a significant contribution of the converse flexoelectricity to the electromechanical coupling. Apparent mechanical stresses were created in the inclusion and its surrounding matrix. The disparity between the compression and tension stresses increased with the amplitude of the applied electric potential. The components of the effective coefficients that couple a polarization gradient and strain in the longitudinal and transverse directions ($\overline{f}_{2222}$ and $\overline{f}_{2211}$) demonstrated an approximately similar tendency for the addressed geometry. The equivalent coefficients of a composite with a 20% volume fraction reached up to 50% the flexo phase. These coefficients steadily increased with the increasing volume fraction, which fits with the rule of mixture. However, the change in the shear constant, $\overline{f}_{1212}$, was found to be marginal. Therefore, further study is essential to evaluate its performance. Additionally, the study examined the impact of material elasticity. A stronger effect was observed when using softer inclusion.

Funding: This research received no external funding

Data Availability Statement: The author declares that the data supporting the findings of this study are available within the paper.

Conflicts of Interest: The author declares no conflicts of interest.

References

1. Fousek, J.; Cross, L.; Litvin, D. Possible piezoelectric composites based on the flexoelectric effect. *Mater. Lett.* **1999**, *39*, 287–291. [CrossRef]
2. Maranganti, R.; Sharma, N.; Sharma, P. Electromechanical coupling in nonpiezoelectric materials due to nanoscale nonlocal size effects: Green's function solutions and embedded inclusions. *Phys. Rev. B* **2006**, *74*, 014110. [CrossRef]
3. Kogan, S.M. Piezoelectric effect during inhomogeneous deformation and acoustic scattering of carriers in crystals. *Sov. Phys. Solid State* **1964**, *5*, 2069–2070.
4. Tagantsev, A.K. Theory of flexoelectric effect in crystals. *Sov. Phys. JETP* **1985**, *61*, 1246–1285.
5. Zubko, P.; Catalan, G.; Buckley, A.; Welche, P.; Scott, J. Strain-gradient-induced polarization in SrTiO 3 single crystals. *Phys. Rev. Lett.* **2007**, *99*, 167601. [CrossRef]
6. Shen, S.; Hu, S. A theory of flexoelectricity with surface effect for elastic dielectrics. *J. Mech. Phys. Solids* **2010**, *58*, 665–677. [CrossRef]
7. Wang, B.; Gu, Y.; Zhang, S.; Chen, L.Q. Flexoelectricity in solids: Progress, challenges, and perspectives. *Prog. Mater. Sci.* **2019**, *106*, 100570. [CrossRef]
8. Ma, W.; Cross, L.E. Flexoelectricity of barium titanate. *Appl. Phys. Lett.* **2006**, *88*, 232902. [CrossRef]
9. Ma, W.; Cross, L.E. Flexoelectric polarization of barium strontium titanate in the paraelectric state. *Appl. Phys. Lett.* **2002**, *81*, 3440–3442. [CrossRef]
10. Fu, J.Y.; Zhu, W.; Li, N.; Cross, L.E. Experimental studies of the converse flexoelectric effect induced by inhomogeneous electric field in a barium strontium titanate composition. *J. Appl. Phys.* **2006**, *100*, 024112. [CrossRef]
11. Shu, L.; Huang, W.; Ryung Kwon, S.; Wang, Z.; Li, F.; Wei, X.; Zhang, S.; Lanagan, M.; Yao, X.; Jiang, X. Converse flexoelectric coefficient f_{1212} in bulk $Ba_{0.67}Sr_{0.33}TiO_3$. *Appl. Phys. Lett.* **2014**, *104*, 232902. [CrossRef]
12. Abdollahi, A.; Domingo, N.; Arias, I.; Catalan, G. Converse flexoelectricity yields large piezoresponse force microscopy signals in non-piezoelectric materials. *Nat. Commun.* **2019**, *10*, 1266. [CrossRef]

13. Zhang, S.; Liu, K.; Wen, X.; Wu, T.; Xu, M.; Shen, S. Converse flexoelectricity with relative permittivity gradient. *Appl. Phys. Lett.* **2019**, *114*, 052903. [CrossRef]
14. Wang, Y.; Tang, Y.; Zhu, Y.; Feng, Y.; Ma, X. Converse flexoelectricity around ferroelectric domain walls. *Acta Mater.* **2020**, *191*, 158–165. [CrossRef]
15. Li, Q.; Nelson, C.; Hsu, S.L.; Damodaran, A.; Li, L.L.; Yadav, A.; McCarter, M.; Martin, L.; Ramesh, R.; Kalinin, S. Quantification of flexoelectricity in PbTiO3/SrTiO3 superlattice polar vortices using machine learning and phase-field modeling. *Nat. Commun.* **2017**, *8*, 1468. [CrossRef] [PubMed]
16. Maranganti, R.; Sharma, P. Atomistic determination of flexoelectric properties of crystalline dielectrics. *Phys. Rev. B* **2009**, *80*, 054109. [CrossRef]
17. Hong, J.; Vanderbilt, D. First-principles theory and calculation of flexoelectricity. *Phys. Rev. B* **2013**, *88*, 174107. [CrossRef]
18. Tian, X.; Xu, M.; Deng, Q.; Sladek, J.; Sladek, V.; Repka, M.; Li, Q. Size-dependent direct and converse flexoelectricity around a micro-hole. *Acta Mech.* **2020**, *231*, 4851–4865. [CrossRef]
19. Sidhardh, S.; Ray, M. Exact solutions for flexoelectric response in elastic dielectric nanobeams considering generalized constitutive gradient theories. *Int. J. Mech. Mater. Des.* **2019**, *15*, 427–446. [CrossRef]
20. Liu, C.; Wang, J. Size-dependent electromechanical properties in piezoelectric superlattices due to flexoelectric effect. *Theor. Appl. Mech. Lett.* **2017**, *7*, 88–92. [CrossRef]
21. Sharma, N.; Maranganti, R.; Sharma, P. On the possibility of piezoelectric nanocomposites without using piezoelectric materials. *J. Mech. Phys. Solids* **2007**, *55*, 2328–2350. [CrossRef]
22. Schröder, J.; Keip, M.A. Two-scale homogenization of electromechanically coupled boundary value problems. *Comput. Mech.* **2012**, *50*, 229–244. [CrossRef]
23. Maruccio, C.; De Lorenzis, L.; Persano, L.; Pisignano, D. Computational homogenization of fibrous piezoelectric materials. *Comput. Mech.* **2015**, *55*, 983–998. [CrossRef]
24. Singh, D.; Sharma, S.; Kumar, R.; Vaish, R. Representative volume element model of triply periodic minimal surfaces (TPMS)-based electrostrictive composites for numerical evaluation of effective properties. *Acta Mech.* **2022**, 234, 355–375. [CrossRef]
25. Rahmati, A.H.; Liu, L.; Sharma, P. Homogenization of electrets with ellipsoidal microstructure and pathways for designing piezoelectricity in soft materials. *Mech. Mater.* **2022**, *173*, 104420. [CrossRef]
26. Mohammadi, P.; Liu, L.; Sharma, P. A theory of flexoelectric membranes and effective properties of heterogeneous membranes. *J. Appl. Mech.* **2014**, *81*, 011007. [CrossRef]
27. Sidhardh, S.; Ray, M. Effective properties of flexoelectric fiber-reinforced nanocomposite. *Mater. Today Commun.* **2018**, *17*, 114–123. [CrossRef]
28. Guinovart-Sanjuán, D.; Merodio, J.; López-Realpozo, J.C.; Vajravelu, K.; Rodríguez-Ramos, R.; Guinovart-Díaz, R.; Bravo-Castillero, J.; Sabina, F.J. Asymptotic homogenization applied to flexoelectric rods. *Materials* **2019**, *12*, 232. [CrossRef] [PubMed]
29. Kalamkarov, A.; Hadjiloizi, D.; Weaver, P.; Georgiades, A. Asymptotic homogenization of flexoelectric composite plates with periodically varying thickness. *Math. Mech. Solids* **2022**, 29, 10812865221136269. [CrossRef]
30. Kundalwal, S.; Shingare, K.; Rathi, A. Effect of flexoelectricity on the electromechanical response of graphene nanocomposite beam. *Int. J. Mech. Mater. Des.* **2019**, *15*, 447–470. [CrossRef]
31. Shingare, K.; Kundalwal, S. Flexoelectric and surface effects on the electromechanical behavior of graphene-based nanobeams. *Appl. Math. Model.* **2020**, *81*, 70–91. [CrossRef]
32. Shingare, K.; Naskar, S. Probing the prediction of effective properties for composite materials. *Eur. J. Mech.-A/Solids* **2021**, *87*, 104228. [CrossRef]
33. Yvonnet, J.; Chen, X.; Sharma, P. Apparent flexoelectricity due to heterogeneous piezoelectricity. *J. Appl. Mech.* **2020**, *87*, 111003. [CrossRef]
34. Mawassy, N.; Reda, H.; Ganghoffer, J.F.; Eremeyev, V.A.; Lakiss, H. A variational approach of homogenization of piezoelectric composites towards piezoelectric and flexoelectric effective media. *Int. J. Eng. Sci.* **2021**, *158*, 103410. [CrossRef]
35. Hamdia, K.M. A representative volume element model to evaluate the effective properties of flexoelectric nanocomposite. *Eur. J. Mech.-A/Solids* **2024**, *103*, 105149. [CrossRef]
36. Barceló-Mercader, J.; Codony, D.; Mocci, A.; Arias, I. Computational homogenization of higher-order electro-mechanical materials with built-in generalized periodicity conditions. *Comput. Methods Appl. Mech. Eng.* **2024**, *423*, 116861. [CrossRef]
37. Chen, X.; Yvonnet, J.; Park, H.; Yao, S. Enhanced converse flexoelectricity in piezoelectric composites by coupling topology optimization with homogenization. *J. Appl. Phys.* **2021**, *129*, 245104. [CrossRef]
38. Shu, L.; Wei, X.; Pang, T.; Yao, X.; Wang, C. Symmetry of flexoelectric coefficients in crystalline medium. *J. Appl. Phys.* **2011**, *110*, 104106. [CrossRef]
39. Gao, X.L.; Park, S. Variational formulation of a simplified strain gradient elasticity theory and its application to a pressurized thick-walled cylinder problem. *Int. J. Solids Struct.* **2007**, *44*, 7486–7499. [CrossRef]

40. Ghasemi, H.; Park, H.S.; Rabczuk, T. A level-set based IGA formulation for topology optimization of flexoelectric materials. *Comput. Methods Appl. Mech. Eng.* **2017**, *313*, 239–258. [CrossRef]
41. Abdollahi, A.; Peco, C.; Millan, D.; Arroyo, M.; Arias, I. Computational evaluation of the flexoelectric effect in dielectric solids. *J. Appl. Phys.* **2014**, *116*, 093502. [CrossRef]
42. Hughes, T.J.; Cottrell, J.A.; Bazilevs, Y. Isogeometric analysis: CAD, finite elements, NURBS, exact geometry and mesh refinement. *Comput. Methods Appl. Mech. Eng.* **2005**, *194*, 4135–4195. [CrossRef]
43. Ghasemi, H.; Park, H.S.; Zhuang, X.; Rabczuk, T. Three-dimensional isogeometric analysis of flexoelectricity with MATLAB implementation. *Comput. Mater. Contin.* **2020**, *65*, 1157–1179. [CrossRef]

MDPI AG
Grosspeteranlage 5
4052 Basel
Switzerland
Tel.: +41 61 683 77 34

Computation Editorial Office
E-mail: computation@mdpi.com
www.mdpi.com/journal/computation

www.ingramcontent.com/pod-product-compliance
Lightning Source LLC
Chambersburg PA
CBHW061553210926
51850CB00009B/2093